D1290158

FINITE ELEMENT PRIMER

ELLIS HORWOOD SERIES IN ENGINEERING SCIENCE

FINITE ELEMENT PRIMER

BRUCE IRONS, B.Sc., D.Sc.
Professor of Civil Engineering
University of Calgary, Alberta, Canada

and

NIGEL SHRIVE, M.A., D.Phil.
Associate Professor of Civil Engineering
University of Calgary, Alberta, Canada

ELLIS HORWOOD LIMITED
Publishers · Chichester

Halsted Press: a division of
JOHN WILEY & SONS
New York · Brisbane · Chichester · Toronto

First published in 1983 by
ELLIS HORWOOD LIMITED
Market Cross House, Cooper Street, Chichester, West Sussex, PO19 1EB, England

The publisher's colophon is reproduced from James Gillison's drawing of the ancient Market Cross, Chichester.

Distributors:

Australia, New Zealand, South-east Asia:
Jacaranda-Wiley Ltd., Jacaranda Press,
JOHN WILEY & SONS INC.,
G.P.O. Box 859, Brisbane, Queensland 40001, Australia

Canada:
JOHN WILEY & SONS CANADA LIMITED
22 Worcester Road, Rexdale, Ontario, Canada.

Europe, Africa:
JOHN WILEY & SONS LIMITED
Baffins Lane, Chichester, West Sussex, England.

North and South America and the rest of the world:
Halsted Press: a division of
JOHN WILEY & SONS
605 Third Avenue, New York, N.Y. 10016, U.S.A.

© 1983 B. Irons and N. Shrive/Ellis Horwood Ltd.

British Library Cataloguing in Publication Data
Irons, Bruce
Finite element primer. –
(Ellis Horwood series in engineering science)
1. Finite element method
I. Title   II. Shrive, Nigel
515.3'53   TA347.F5

Library of Congress Card No. 82-23392

ISBN 0-85312-271-7 (Ellis Horwood Ltd., Publishers – Library Edn.)
ISBN 0-85312-440-X (Ellis Horwood Ltd., Publishers – Student Edn.)
ISBN 0-470-27414-7 (Halsted Press – Library Edn.)
ISBN 0-470-27415-8 (Halsted Press – Student Edn.)

Typeset in Press Roman by Ellis Horwood Ltd.
Printed in Great Britain by R. J. Acford, Chichester

Summary of Contents

Authors' Preface

This text is for engineers. We believe that engineers want to know four things about finite elements:

(a) What they are
(b) How they work
(c) How to use them
(d) What can go wrong.

We hope this Primer will 'prime' you adequately.

Thanks

Carol Irons
Sue Shrive
Ellis Horwood (publisher)
Denise Hanoski (typist)
Fred Sowan (house editor)
Mick Wasley (typesetter)
Bob Harris (paste-up)

Apologia [†]

Authors' Preface

If you are past a certain age, you may be vexed by what we have done to the English language. We simply try to write as we would speak to a class, or to an individual student. We *can* write formally, if our arms are twisted enough. Otherwise we wouldn't have doctorates. But formal technical English is *so* boring.

When you go out to work you may *have* to write formally. Some firms are adventurous, others are old-fashioned. How engineers will write, twenty years from now, is anyone's guess. There are even suggestions that you will have to use a sort of Pidgin English, with only 800 words, so that some fool computer can translate your reports into foreign languages. Doesn't that sound horrible? Even worse than formal English.

[†] We aren't ashamed of anything. "Apologia" means an explanation of what we are doing, and why.

List of Symbols

GREEK SYMBOLS

ϵ, γ: strains, typically ∂(deflection)/∂(initial coordinate)
   ϵ, extensional; γ, shear.
σ, τ: stresses, typically force per unit area
   σ, normal; τ, shear. The subscripts used in the text denote firstly
   the direction of the normal to the face on which the force acts,
   and secondly the direction of the force.
ξ, η: local coordinates for an isoparametric quadrilateral, in general
   both curvilinear and non-orthogonal.
θ: rotation.
ν: Poisson's ratio, how much the diameter of a wire decreases when
   we stretch it.

SYMBOLS, CONVENTIONS, and DEFINITIONS

Assembly of elements: See *Front solvers, Destination vector*. Given
   the element stiffnesses and load contributions, how to assemble
   the large set of equations to be solved.
B: Strain-displacement matrix. A term is

$$\frac{\partial \text{ (strain)}}{\partial \text{ (nodal variable, normally a deflection)}}$$

One column is for each nodal variable. 'Strain' is generalized
   here, to mean anything that generates strain energy directly, at
   the infinitesimal level.
Band solvers: These take advantage of the fact that the assembled
   stiffness matrix is usually well banded. The user must aim to

minimize, in every element, (largest node number) minus (smallest node number) because this controls the amount of storage needed. See *Front solvers*.

Bandwidth: The maximum spread of columns with numbers in any row of a matrix. e.g.: in Table 4.1 the bandwidth is 8.

Code: (verb) A jargonistic term implying the ability both to think out the logic for and to write a computer program.

Conditioning, Numerical conditioning: A job is 'ill-conditioned' if the results that interest us are abnormally sensitive to roundoff.

Deflection: How far something moves, maybe in the x-direction, etc. See *Displacement*.

Destination vector: A sequence of row or column addresses for the element variables in turn, in (a) the very large assembled matrix. However, this is much too large in practice. Thus (b) in real programs the vector gives the addresses in the smaller submatrix relating to the variables in the current 'front', which are 'active' at present in the reduction process.

Displacement: Strictly, displacement is the vector change in position of a point defined within a body, when the body moves and/or deforms. 'Deflection' (q.v.) means a component of displacement. Both are often used loosely, however.

E: Young's modulus, which relates tensile stress to strain for a wire.

Earthing: A point rigidly fixed to the gound so that it does not move in one or more directions.

Eigenvalues: see *Positive Definite,* also section 6.12.

Femski: Finite Element Student Key-Instructor: Our finite element scheme. You may have it or something similar.

Fixities: A point fixed to earth (see *Earthing*) with zero displacement in one or more directions.

Flying structure: A structure inadequately supported, a particular case of a 'mechanism'.

Front solvers: These use the better technique of housekeeping on the computer, from most points of view. When for example the solver has accepted the first six out of twenty elements, the 'front' comprises the nodes along the boundary between the substructure comprising elements 1–5, and the substructure comprising elements 6–20, plus in most existing programs the remaining nodes in element 6. We try to order the elements so as to minimize the 'front' at all times. See *Band solvers*.

G: Shear modulus = d(shear stress)/d(shear strain).

Hoop stresses: Tensile stress in a pipe or sphere with internal pressure.

Hybrid element: One derived by an advanced technique, in which the stress patterns are guessed, rather than the displacement patterns. Done well, this gives greatly improved performance. (See Appendix.)

Hydraulic jack: A device used in structural laboratories to create large loads. Oil is forced into a cylinder, firmly bolted to earth. The pressure operates on a piston, which applies the load to the structure being tested.

Integrand: The function to be integrated, what lies between the integral sign $\int$ and dx, etc.

Inverse matrix: For example,

$$\begin{bmatrix} 3 & 0 & 2 \\ 7 & 1 & 3 \\ 1 & 0 & 1 \end{bmatrix} \begin{bmatrix} 1 & 0 & -2 \\ -4 & 1 & 5 \\ -1 & 0 & 3 \end{bmatrix} = \begin{bmatrix} 1 & 0 & 0 \\ 0 & 1 & 0 \\ 0 & 0 & 1 \end{bmatrix}$$

The matrix on the right is known as the 'identity matrix' **I**, because **IA = A** and **BI = B**. (Think about ¼ × 4 = 1). The other two matrices: each is the 'inverse' of the other.

Isoparametric element: The formal definition is in the footnote to section 3.7. A friendlier definition follows. Consider an element in the $\xi-\eta$ field, physically. Give it very large deflections, to bring the nodes to the $(x_i \ y_i)$ positions. Now regard this as the starting condition.

Isotropic: The same elastic properties in any direction.

Jack: A hydraulic jack.

Jacobian: A matrix relating local coordinates to global coordinates.

k: Stiffness or 'rate' of a simple spring, d(load)/d(deflection).

K: Stiffness matrix, square, positive definite (q.v.). A term is ∂(load component)/∂(displacement component). (See *Displacement, Deflection*.)

L: Weight per unit area of a spring-supported membrane.

Local coordinates: Coordinates peculiar to an element, as against the 'global coordinates', *x, y, z*.

Matrices: Arrays of numbers. To multiply two matrices, take rows into columns:

$$\begin{bmatrix} a & b & c \\ d & e & f \\ g & h & i \end{bmatrix} \begin{bmatrix} j & m \\ k & n \\ l & p \end{bmatrix} = \begin{bmatrix} aj + bk + cl, & am + bn + cp \\ dj + ek + fl, & dm + en + fp \\ gj + hk + il, & gm + hn + ip \end{bmatrix}$$

Note: **A** would be a matrix

A would be a single number.

Mechanism: A structure which, for certain patterns of deformation, generates no strain energy. Hence it cannot support loads, and the assembled equations cannot be solved.

Mesh: The outlines of the elements used to model the object of interest. The outline of the mesh should give the appropriate view of the object being modelled.

N: Shape function matrix. A term is

$$\frac{\partial \text{ (displacement component at arbitrary point)}}{\partial \text{ (nodal variable, normally a deflection)}}$$

One column is for each nodal variable. Any column is the displacement vector if that variable is made unity.

Nodal valency: see *Valency*.

Node: A point in space, where two or more elements are 'joined' by sharing the same variable(s).

Numerical conditioning: see *Conditioning*.

Numerical integration: Like Simpson's rule, a technique for estimating an integral by putting values at certain points within the domain into a given formula.

p: Pressure, in the same units as the stresses, σ and τ.

Patch test: A simple test on the computer as to whether the solution with a given element will converge towards the right answers, as the mesh is refined indefinitely.

Pivots: In reducing, prior to solving equations, the value of a coefficient just before we divide by it. With N equations there are only N pivots. If they are all positive, the matrix is positive definite, and diagonal pivots must be used, which preserves symmetry. The pivots are the numbers lying just above one step in Crout's tabular layout (e.g.: 0.5, 0.444, 0.484, etc. in Table 2.3). If any pivot is zero or negative, give up — the devil got there first!

Poisson's ratio: A measure of how much a wire shrinks in cross-section when we stretch it.

Positive definite: If a structure is in equilibrium, any small perturbation requires a second-order amount of work (see *Virtual work*) but necessarily *positive*. In matrix notation, the quadratic form, e.g. $V^T K V$, is greater than zero for any non-null vector V. In numerical terms, when K is reduced as in solving equations, the pivots (q.v.) are all positive. The eigenvalues (q.v.) are also positive.

Prescribed variables: These have values that are fixed in advance, so that if the nth variable is prescribed we ignore the nth equation.

Property list (Element): Numbers 1 to N in sequence. In FEMSKI you have to state this number, when you enter the node numbers for an element. Later, you have to state the properties themselves. The first 'property' is the element *type,* e.g.: plane stress, beam. There follows a list, like Young's modulus, density, temperature .... But don't worry. FEMSKI will ask all the right questions. And another scheme is likely to be altogether different.

Reactions: Forces which support the structure, against gravity and other applied loads.

Redundancy: A structure N times redundant continues to support any load if N carefully chosen members (simple springs, ties, struts, etc.) are removed.

Residuals: Suppose we are solving

$$x^2 + 3x = 2$$

and we get $x = 0.56$. Putting this value into the equation gives 1.9936 instead of 2. Our answer is wrong, and the 'residual' is -0.0064.

Roundoff: The consequences, for example, of only being able to store 12 digits in all, for a decimal number. Computers and calculators only store and manipulate numbers to a certain number of decimal places. Some truncate, that is, they subtract the later decimals that are not used. The better systems round off to the nearest decimal, or at least attempt to increase numbers as much as they reduce them, on average.

Scheme: A computer program containing finite elements, their solution and the input/output packages.

Shape function constraints: Conceptual rigid additional parts, which would constrain a real structure to deform as permitted by the finite element assumptions.

Shape function: The deflections, etc., throughout an element, if one and only one variable is given the value unity.

Shear: Deforming, without change of volume.

Simple structure: Every member is necessary. Add another member and it becomes 'redundant' (q.v.). Remove any member and it becomes a 'mechanism' (q.v.) and collapses.

Solvers: See *Band solvers, Front solvers*.

Spring to earth: A stiffness at a node in a particular 'direction': the node can be free in other 'directions'. The 'direction' could be a displacement or a rotation. One end of the spring is attached to the node, the other to the earth (fixed). Thus an infinitely stiff spring fixes (earths) the node.

Stiff: One structure, or representation of a structure, is 'stiffer' than another if for *any* given loads the strain energy is less.

Strain energy: For example, the work stored in a spring.

Stress matrix: (3 × 3) A useful artifice for combining all six stresses into one entity, in such a way that we can perform meaningful mathematical operations with it, and learn more about the stress systems. Alias *Stress tensor*.

T: Tension in a wire.

Tensile: Stretching.

Terminal: A sort of typewriter, usually with a TV screen instead of paper, which communicates directly with a computer: like a card-punch but without cards.

u, v, w: Deflections, which see. w = vertical deflection.

$[u, v, w]$: Displacement vector, change in position, u = change in x, etc.

Valency, Nodal valency: The number of elements that share a node.

Variables: u, v and w deflections in the x, y and z directions, usually but not always. Sometimes rotations are variables, etc.

Vertices: The corners of a triangle, quadrilateral, etc.

Virtual work: If we have equilibrium, and we compress a spring slightly further, then the work calculated, assuming that the load does not change, equals the increase in strain energy, q.v.

Warping: A section of a beam (prism), initially plane, deforms out of plane under load.

Whirling speed: When rotating machinery reaches a certain r.p.m. the slightest unbalance causes a violent vibration. (This happens even at the relatively low r.p.m. of a spin-dryer.)

X: External loads at nodes.

x, y, z: Cartesian coordinates of a point: x, y, z axes must be orthogonal.

Young's modulus: A measure of how stiff a material is in tension.

CHAPTER 1

How are finite elements used?

1.1 HI!

Wherever we load something solid, and if there is an opportunity of improving the design, then somebody, somewhere, is attempting to do so using finite elements. Those industries with a long tradition of careful stressing — aircraft, bridges, arch dams, buildings etc. — are already spending billions on finite element computer programs. But also cars, soils, ships, medicine, metal forming. . .

We can calculate stresses, and deflections too, and vibrations due to out-of-balance or flutter, shaking due to earthquakes, wind, waves bombs etc. Then there are more speculative ventures: heat flow is common, but what of the flow of water, oil and blood, pollution control, electromagnetism ...? At the time of writing, finite elements often feature in job advertisements, and no engineer can *afford* to remain totally ignorant!

Why 'finite elements'? Think (if you can bear to) about two other courses. In 'Strength of Materials' we write and solve differential equations; that is, we base our work on the *infinitesimal* concepts of stress and strain. Something stretches. That is, the atoms are drawn further apart than they want to be. We treat it as a continuum. No atoms, continuous material. Then we apply differential calculus, to find the stretch *at a point*. Like speed, at an instant of time. This is really an assumption, a *conceptual model*. Not that it will ever make any difference, even with the tiniest microbe. Remember, models. Little lies.

Engineering is full of little lies. In 'Structures', we do the opposite, in a sense. We have a complicated mass of vaguely rectangular objects, all welded together. We say, treat this as a beam, that as a tie, it is rigidly supported by the earth, there. All untrue, but near enough.

We take simple members, like beams, ties, struts, and perhaps lumps stiff enough to be called 'rigid'. Then we connect them all together, so as to maximize the stress of the examination. . .

Now think of a car body. There are clearly beams, but there are also complex shells, and welds, and doors, and machinery, and occupants. . . crash simulation will require an assortment of 'elements' of *finite* size — hence the name, 'finite element'. One job may use several types of element.

The individual elements are more complicated than beams, but most are simpler than our 'Strength of Materials' problems. There is a wide variety. We may use ten elements in one job, thousands in another. Two people will model the same job differently.

We can mix big and small elements, because they are of fairly general shape, like quadrilaterals. We use lots of tiny elements, for intricate shapes; also where the stresses are high, or where they vary rapidly. If we do not like the answers, we run the job again using smaller elements, and more of them.

You knew nothing about finite elements before this course started? You were unaware that they are a billion-dollar industry?† This would not happen merely because a few grey-haired professors find them amusing. Obviously, finite elements work. But the next four chapters may convey the opposite impression. In developing our explanation of how finite elements work, we collect together various types of *misbehaviour* that we know, and then simplify the examples so that the reasons become obvious. A strange way to teach? This is how one frequently learns, especially in industry. The experts there have learnt much of what *they* know, from commercial jobs that have gone wrong! At one stage we intended to call this book 'Games that Finite Elements Play'. The idea is to win, mostly.

1.2 HI, FEMSKI!

Now this book is written throughout on the assumption that your teacher has our program FEMSKI. If not, you needn't fret. There are many good finite element schemes available; besides, the one chosen by your teacher may be already running on your computer. For example, scheme-writers are very busy producing schemes for micro-computers. Of course, we would be happier if you are using FEMSKI, because some features seem to be unique, and because it is all yours.

---

† Not quite: 0.6 billion turnover (1979) and growing (USA billion, 10^9. Ed.).

We give it away, and — unlike most commercial vendors — we give you the Fortran source code, so that you or your teacher can alter it, if you dare to. It does contain some unusual elements. But some of the newer schemes also have interesting elements, and other useful features too. Most important, FEMSKI was developed from the program that is the centrepiece of the advanced book 'Techniques of Finite Elements', so that if you choose this route to advance your education, most things will fit into place fairly easily. If your teacher has selected another scheme, he should be able to find elements that more or less match those that we have used in writing this book.

Do you want to see a few successful examples, before we shatter your confidence? Throughout this course you will have assignments on the computer, in addition to the theory — lots of them. We insist that you do them. You will benefit much more if you can do the practical work as well as the theory. But it should be fun.

In Fig. 1.1 a thin steel sheet is loaded in its own plane. The four 'elements' marked I–IV comprise what is called the 'mesh'. When the cantilever deforms, the 'nodes' move. The nodes are the corners marked 1–10 in this case. The model assumes there are two unknown deflections at each node, one in each of the x- and y-directions: $u_1, v_1, u_2 \ldots v_{10}$ at nodes 1–10, twenty variables in all. For these twenty variables the computer must generate twenty equations and

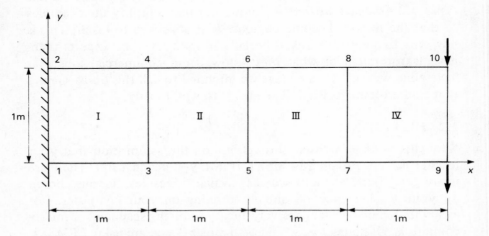

Fig. 1.1 — A finite element representation of a cantilever, shown in side elevation, with a vertical endload. The Roman numerals are element numbers. The ordinary numbers represent nodes, or points where the deflections in the x and y directions are the variables.

solve them. The coefficients of $u_1 \ldots v_{10}$ on the left-hand sides of the equations are called 'stiffnesses', by analogy with a simple spring. That is, Stiffness × Deflection = Force, $kd = F$. The right-hand sides are the applied loads. People call them 'right-hand sides', or 'load cases'.

Twenty equations. But we lose four of them, when we force u_1, v_1, u_2 and v_2 to be zero. In effect, we invoke the reactions from earth. Observe what the variables imply: the elements really *join* at nodes 3–8. For example u_5 is the same in element II and in element III. Of course: the material is continuous across element boundaries!

Here each element is an identical square, but our scheme FEMSKI (Finite Element Method, Student Key-Instructor) would have allowed four different quadrilateral shapes. Even in a simple job like this, a large commercial scheme would also allow a choice of several element formulations. This is hardly surprising – there might be as many as 400 000 FORTRAN statements! Even with little FEMSKI you have a choice of two. The better is a 'hybrid' formulation. Although the theory is well beyond the scope of this book, the element is quite easy to use. Having selected this element, of type 8, and having drawn the mesh for your job as in Fig. 1.1, you must punch the cards, which explain the details to the computer, in terms that any such idiot can understand. . .

1.3 HI, COMPUTER!

Or better, we hope that somebody will buy you a console ('terminal' in America). You will press more or less the same buttons as on a card-punch. But a terminal keeps asking you for what it wants next, which really helps. Amusing, to meet a chatty computer. You begin by 'logging-in'; that is, you tell him your secret code-name, and he tells you his – that protects you both against impostors! Then you must call the program, by typing maybe 'ec prep'. Your computer soon responds, at length:

'**Greetings. First, a good title for your job please** –'
'**Start typing** –' Cantilever with endload, four hybrid elements.
'**Now for each element, the property-list number, followed by the node numbers in anticlockwise order** – **one line per element. (For the last element, make the properties number negative.)**'
'**Please use fixed format here. I shall write 'u' to indicate the unit-columns.**'

He waits (patiently) for you to answer:

u	u	u	u	u
1	1	3	4	2
1	3	5	6	4
1	5	7	8	6
−1	7	9	10	8

Node-numbers of elements I–IV; anticlockwise order (you can start at any node in the element).

Property-list number.

Actually he helps, a little. He writes 'Element 1' before you type the first line, and so on. Note the '−1', which informs him that (7,9,10,8), element IV, is the last. If you forget the minus sign, don't panic. Type 'minus' (−) as the only character on the *next* line. Either way, he deduces correctly that you have just four elements, all of the same material.[†]

So he quizzes you further:

'Element property list 1: what TYPE NUMBER had you in mind?'

You reply '8', and he comments:

'That's the super-duper hybrid 4-node quad?'

Without waiting for a reply he goes on to enquire:

'What is Young's modulus?'

Tricky question. You might enter 210000000000. = E for steel, in N/m^2. Or better, 2.1e11 which means the same, 2.1×10^{11}, as any book on FORTRAN will tell you. Young's modulus is explained in Chapter 6.

Or perhaps he gave you an element description that you did not expect. So you would like to correct the '8'? You'd give a million bucks? Well, not quite. Type 999999 as 'Young's modulus'. He will go back, and ask you again for the element type. Try harder next time.

But let us assume '8' was correct. His next question:

'What is Poisson's ratio?'

You must tell him the value for steel, 0.3. Next he asks for:

'The DENSITY of the material?'

[†] If you are very lazy, you may omit the '1' in the first three lines. He reads 'zero' and PREP changes this to '1'.

This is 76300. in N/m^3. You have to use the *force* per unit volume, because the stresses will be in Pascals, that is, N/m^2. Remember the decimal point.

Finally he asks for the thickness. We must explain something. To enter one centimeter, you would write 0.01 (metres), but *zero* implies an altogether different case, known as 'plane strain'. See Chapter 6.

In entering the material properties, make a habit of *bunching the numbers together at the left,* without any spaces. In FEMSKI, 210000000000. would not be acceptable. You *must* use 2.1e11, because it has to fit into the first ten columns. And 999999 or 'minus' will only work if there are no spaces.

This completes the material properties. (He remembers that you only mentioned the one list, in your first four 'cards'). Next, he instructs you:

'Now write some of the node numbers, followed by the fixing code (zero if the node is free) and by the coordinates, x, y or x, y, z.

'Note, you must present the last node number as negative.'

('Fixing codes'. Suppose there are three variables at the node in question, u, v and w. Then 111 would mean that all three are zero, 100 that only u is fixed, 11 that v and w are both earthed. Interpret 11 as 011, with zero in the 'hundreds' column, for u. These codes are read as ordinary decimal numbers, but FEMSKI interprets them as multiple on-off switches.)

Your reply:

Node Number	Fixing code	Coordinates x	y	z
1	11	0.0	0.0	
2	11	0.0	0.1	
3	0	1.0	0.0	
4	0	1.0	1.0	
5	0	2.0	0.0	
6	0	2.0	1.0	
7	0	3.0	0.0	
8	0	3.0	1.0	
9	0	4.0	0.0	
−10	0	4.0	1.0	

He types these two lines for you.

The fixing code '11' means that u and v are fixed as in Fig. 1.1. Remember here there are only these two variables.

'0' means that the deflections u and v are both free.

Don't waste time making neat columns here. But you must leave spaces between the numbers!

The −10 means that the list is complete. As with the material property lists, this is all in 'free format'. Most computers can accept 27.0, 27., 2.7e1, for example, or any mixture of these styles.

We have not quite finished yet. In fact there could be a lot more input. He goes on:

> 'Is that all the FIXITIES? Are there any NONZERO prescribed values?
> 'If so, type the node, the fixing code, and the values, including zeros for those that aren't fixed.
> 'Tell me when you've finished, with a negative node-number.'

Alas, you have to disappoint him. Type a minus sign (−), nothing else, and the carriage return as usual. He continues:

> 'Have you any EXTRA LOADS, that I don't know about? Give me the node-numbers, followed by the nodal loads in turn.
> 'End with a negative node-number.'

He already has the gravity-loads. You gave him the density, in force-units. But there are additional loads, at nodes 9 and 10, much larger:

$$9 \quad 0. \quad 100000.$$
$$10 \quad 0. \quad 100000.$$

These are vertical loads (Fig. 1.1), i.e. the second variable at nodes 9 and 10. The horizontal loads are zero. All done now, but he still natters on:

> 'Finally, I can accept extra SPRINGS TO EARTH. Give me their stiffnesses, preceded by the node-number.
> 'Signal the last node with a minus sign.'

You signal as before (−), and at last he surrenders:

> 'Thanks, all done − see the data file.' (Table 1.1 displayed.)
> 'Rather quit now? Edit input file? yes or no?' no
> 'Run immediately, while you wait? yes or no?' yes
> 'Cheaper as absentee! Second thoughts?' no.

1.4 HI AGAIN!

Easy. But most people make mistakes, sometimes. In that case, the computer will do his best to find the mistakes for you. Maybe as a beginner, you should re-type the lot. As you get more experienced (at making mistakes) you will discover from the detailed documentation that several techniques are available, for correcting a mistake that you notice as you are typing – we saw an example of the '999999' technique, and there is an '888888' technique for re-typing the line where you have just goofed. It would be tedious to over-emphasize this aspect, and perhaps a little premature too, when you haven't made even your *first* mistake yet.

FEMSKI has an attractive feature, however, which provides you with the ultimate in mistake-correctors. As you typed, 'prep' was creating a *data file* in the computer, shown in Table 1.1. See how the words make everything clear; now you can work out, which number is wrong. Do you remember your FORTRAN course, how you could use the 'editor' to correct your program? Nobody expected you to re-type it all, every time! The editor is the only real advantage of a terminal over a card-punch, isn't it? You can change the data, before it goes into FEMSKI.

Are you still puzzled? We have trained the computer to ignore all the *words* in Table 1.1, and to concentrate on the *numbers*. Of course, to do this we have to put the numbers and the words in exactly the right positions, at least when they are mixed, in the same line. If you think about it, this opens up all sorts of possibilities. Not just error-correction. You can run a job, successfully, then go back and change the properties, etc., and then run it again. Or you can fill out a suitable 'shell' of text, with entirely new numbers. Or you can leave blanks, where the text should go, and bypass 'prep' entirely; with this approach, you would type exactly as if you were punching cards. The fun is, that teacher need never know ....

1.5 HI, TEACHER!

The computer obviously needs all these numbers. There is no 'make-work' data in FEMSKI. A friendly program, in most ways.

More good news. You know it all, already. We shall have interesting experiences, with some pretty outlandish elements, but the computer will always lead you by the hand. There will be nothing fresh to learn about FEMSKI. It might take you many months, to learn how to use a big commercial package; they are designed to do so many different things. FEMSKI was written especially for overworked students.

Table 1.1 – The data stored in file 05, as created by 'prep' ready for FEMSKI. In the third line, the number of fixed nodes includes earthed nodes and those with nonzero fixities. There is one right-hand side, but the equation-solver can handle many simultaneously. Moreover, it can re-solve with different right-hand sides (different loading cases), very cheaply. It would have to do this repeatedly in an 'iterative', trial-and-error process. 'Maximum right-hand sides' = 1 is a promise never to exceed 1. However, 'Iterative solutions' = 0 means that in these jobs we do not intend to iterate. If you correct this table, don't forget, the last line of nodal coordinates must be associated with *minus* its node-number.

How many elements?	How many fix. nodes?	No. loaded nodes?	No. sprung nodes?	No. of R.H. sides?	Maximum R.H. sides?	Iterative solutions?
4	2	2	0	1	1	0

Element	Prop. list	Node numbers			
1	1	1	3	4	2	0	0	0
2	1	3	5	6	4	0	0	0
3	1	5	7	8	6	0	0	0
4	1	7	9	10	8	0	0	0

(Integer here and in next line must be in correct column.)
– That's the super-duper hybrid plane stress/strain quad?

Property list	1
Element type	8

Young's Modulus	Poisson's ratio	Density y-gravity	Membrane thickness
2.1e11	0.3	76300.	0.01

Nodes	Fixing-codes	Nodal coordinates	
1	11	0.0000000	0.0000000
2	11	0.0000000	1.0000000
3	0	1.0000000	0.0000000
4	0	1.0000000	1.0000000
5	0	2.0000000	0.0000000
6	0	2.0000000	1.0000000
7	0	3.0000000	0.0000000
8	0	3.0000000	1.0000000
9	0	4.0000000	0.0000000
–10	0	4.0000000	1.0000000

There are no nonzero fixities in this job.

Node	Additional loads	
9	0.000000	100000.000000
10	0.000000	100000.000000

There are no additional springs to earth.

If you have your first laboratory class this week, then it would be a good idea to run a similar, but different example. If you like, the shape may be more complicated. In fact, you could run a big job, if you find this exciting – although you may need special permission. If you want to try something ambitious, however, you need to be careful. The computer is a mean beast. That is why most people proceed very gently, with small jobs at first.

Even professors make mistakes! Any error, however trivial, can stop a job. If not found, it leads quickly to the madhouse, the tantrum room, or the pub – then all your courses will suffer. FEMSKI should give you a clue, from which you can deduce where your data are wrong. We argue thus: you would not give a high rating to a FORTRAN or BASIC package, if it stopped prematurely, without telling you why, or where. Even undergraduates make mistakes. Occasionally.

Having run your job, you now have the duty of presenting the results, as in any other experiment. All those numbers need a good interpreter. You must tell us not only where the part is most likely to break, but also how the load is carried from one region to another. Two distinct tasks, unfortunately. Nobody has really succeeded in combining them in one picture. Have a good try. Ours is in Fig. 1.2.

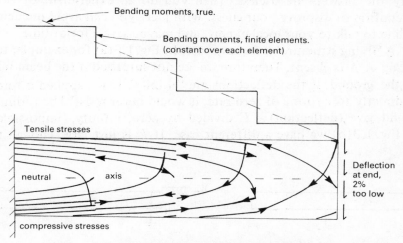

Fig. 1.2 – The cantilever of Fig. 1.1 is modelled with four hybrid 4-node elements. The density of the lines reflects the stress-intensity, as if each line carries the same load. This plot emphasizes the deficiencies: (a) an element can only model constant bending moment, (b) the shear is constant over any section of the beam; hence the fibres or 'stress trajectories' cut the surface, at top and bottom.

Nevertheless, the end-deflection is remarkably good, for such a coarse mesh.

There is another way to earn good marks for your homework. Remember, the computer thrives on good-natured criticism. Estimate the values, by any other means. Not too carefully: but *do it*. For example, you could use your simple beam theory. Then you could add the approximate shear deflections. This is an easy case. As a practising engineer, you will learn to make assumptions, then different assumptions, then look at the side-effects of the assumptions. After five or six very rough sums, you can often 'guestimate' the correct answer. Learn to pit your wits against the computer. In later life, this habit will be your only protection against the evil spirit that lurks inside every computer....

1.6 HI, DEVIL!

One of the best motives for learning about finite elements in some depth is trouble-shooting. You want to understand, quickly, what has gone wrong. Perhaps one day you will not have FEMSKI (or some other scheme with good diagnostics) to look after your interests. FEMSKI will tell you if you forget to earth a job, for example. A 'flying structure'. A dam, or a skyscraper, or an offshore oil-rig. ...

By 'trouble-shooting' we mean that you guess what has happened, why the answers are useless. Then you do an experiment or two, to confirm or disprove your guess. With luck, you can avoid queueing in line to talk to your local expert, and maybe save a lot of time.

A 'flying structure' is an easy case. In Fig. 1.3(a) for example, the spring at A is absent. Therefore we are not surprised if the beam falls to the ground, if the deflections are 'infinite'. If we applied a force F directly to a spring of zero rate, it would never resist. The formula would give (deflection) = F divided by zero, infinity, 'impossible'. In Fig. 1.3(b) we have a different case. If F is just large enough to

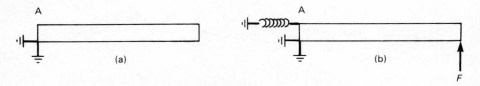

Fig. 1.3 – (a) A cantilever under gravity loads, but we have forgotten to earth it at A. (b) So the spring at A has zero stiffness. Suppose now that the force F exactly balances the weight. The non-existent spring is unloaded. Will it matter now?

bear the weight of the cantilever, the formula gives (zero) ÷ (zero), indeterminate; so that the beam is now in equilibrium for *any* tip deflection, *too* possible — anything will do! Again we cannot expect the computer to solve the equations.

But more often it works out differently. If you do not know what 'roundoff' means, turn to Chapter 8. The spring in Fig. 1.3 is not fanciful. Because the value 'zero' is the result of many intricate calculations, it is seldom *precisely* zero. It might be 10^{-14}. Or it is just as likely to be -10^{-14}. This changes everything. If F does not balance the weight, we get deflections of order 10^{14}. If F is exactly right, we get random numbers, of order 10 times larger than the correct values in practice.

Moreover, the relative deflections in (b) are often surprisingly good. So are the stresses.† Try it and see.

But people like us are *much* too sensible to make such *gross* errors ... . No, most frequently the errors that we and the students make are not funny at all. Stupid mistakes, like mispunched co-ordinates or node numbers, or missing elements: these give crazy stresses, but often only locally. You have to look very carefully every time. Too often, the Devil has the last laugh.

† Very dangerous. We remember a case, admittedly much more involved, where this mistake was discovered long after a paper had been published.

CHAPTER 2

How do finite elements work?

2.1 TEACHER'S PROBLEMS

Chapter 1 was purely descriptive, despite the computer run. This did not really help you to understand finite elements, but we hope it made you feel happier about the course. It is our job to make each step of the theory as easy as possible. This is why our first example will be so trivial. For nearly half the course, we shall concentrate on hand examples, some of which will do the most outrageous and unpleasant things. Very instructive!

We hope that you will trust our conclusions more, because you can check every number with a simple pocket calculator. Because you may find these experiences depressing, however, we maintain the parallel stream, with successful examples run on FEMSKI.

2.2 THE HANGING CABLE

In engineering, we always *adapt* real problems, to suit the techniques at our disposal. The truth is always too complex. We invent a 'model' which introduces *simplifying assumptions*. With finite element models, the assumptions are *physical*. It is one of the real strengths of the method.

Figure 2.1 shows a flexible cable under tension, whose movements are impeded by lengths of closely fitting pipe, one for each element. This is a very simple example of our first general idea, *SHAPE FUNCTION CONSTRAINTS*. The practical question is, whether the modified 'structure' will behave sufficiently like the actual cable, without the tubes. Will the answers be good enough, for the design problem we set ourselves?

Let us see. Let each element be of unit length and of weight L. Thus the cable has to support a vertical force of $\frac{1}{2}L$ at each end of each element, giving $\frac{1}{2}L$ at the nodes A and E, and L at B, C, and D. Because AB $= 1$, the slope of AB equals the deflection w_B, which we assume is small. Because w_B is the same for elements AB and BC, the

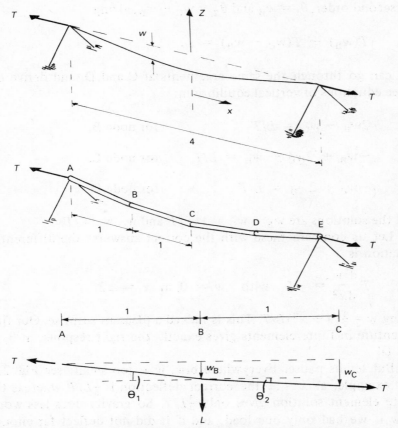

Fig. 2.1 – A cable hangs under its own weight. There are four elements. In each, the deflection w vertically downwards is forced to vary linearly, by the rigid pieces of tube enclosing the cable. The enlarged view at B explains the statics. The angles are small, so that T is constant.

slope of BC is $w_C - w_B$. Or more strictly,

$$\theta_1 = \tan^{-1} w_B, \quad \theta_2 = \tan^{-1}(w_C - w_B).$$

Thus for horizontal equilibrium at B,

$$T_1 \cos \theta_1 = T_2 \cos \theta_2.$$

(The tubes are perfectly lubricated.) Therefore to first order, $T_1 = T_2$, because θ_1 and θ_2 are small, and elsewhere we have simply T, a constant tension.

So the equations for vertical equilibrium are

$$T \sin \theta_1 - T \sin \theta_2 = L.$$

To second order, $\theta_1 = w_B$ and $\theta_2 = w_C - w_B$, giving

$$T(w_B) - T(w_C - w_B) = L.$$

We can go through the same arguments at C and D, and derive the three equations of vertical equilibrium:

$$2w_B - w_C = L/T \qquad \text{for node B,}$$

$$-w_B + 2w_C - w_D = L/T \qquad \text{for node C,}$$

$$-w_C + 2w_D = L/T \qquad \text{for node D,}$$

and the solutions are $w_B = w_D = 1\frac{1}{2}L/T$ and $w_C = 2\,L/T$.

Let us compare these with the correct answers: the differential equation is

$$T\frac{d^2w}{dx^2} = -L \quad \text{with} \quad w = 0 \text{ at } x = \pm 2$$

giving $w = \frac{1}{2}(4 - x^2)L/T$. This is indeed a pleasant surprise. Our first adventure in finite elements gives exactly the right response at B, C, and D!

But let us pause. Everywhere else, w is too small: see Fig. 2.2. For example, at $x = 1\frac{1}{2}$ the correct deflection is $\frac{7}{8}L/T$ whereas the finite element solution gives only $\frac{3}{4}L/T$. So gravity does less work. Now if we had only one load, and if it did not deflect far enough (too little work) then we would say that the modelling of the problem was *TOO STIFF*. In this case also we make the same comment, although we have a distributed load. The rigid tubes *constrain* the structure; it can only deform to certain shapes, which makes it *stiffer*.

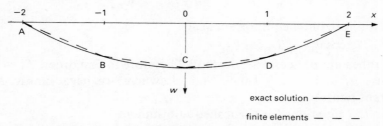

Fig. 2.2 – The finite element solution for Fig. 2.1 compares well with the exact solution. It is unusual however for the answers to be exact at the nodes A ... E.

This always happens, in the strict, law-abiding class of finite elements where continuity is respected, etc.

This is the first general idea. The tubes are called *SHAPE FUNCTION CONSTRAINTS*. They make the real thing behave in a way that the computer can model.

2.3 MEMBRANES IN TENSION, ETC.

The cable problem is admittedly trivial. But the transition to a stretched membrane, with a two-dimensional mesh, is surprisingly direct. Let us consider, not a single horizontal cable, but many, criss-crossing in the x- and y-directions and loaded vertically as in Fig. 2.3(a). Such a mesh of cables can function tolerably well as a membrane — apart from a rather regrettable tendency to leak if it rains.

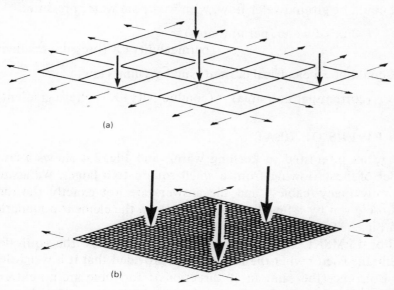

Fig. 2.3 – In (a) a flexible membrane is modelled as a fishnet. The membrane does not exist between the strings. Therefore we must shift the loads to the corners, or at least to the edges. In (b) we fill in each square with countless fine strings. Each is controlled by a rigid tube, so that local denting is impossible.

In (b) we add very many extra cables within each element, so that even with a coarse mesh of elements it becomes possible to accept a load anywhere. Then to every cable we add the shape function constraints. Each is a rigid tube as before, spanning a single element.

Finally, for critics who point out that our model still leaks, we return to the original membrane. Instead of tubes, and strings, we cast infinitely many rigid needles into a thin sheet of rubber. Our picture is now complete. We have the original membrane, and as before we add to it the shape function constraints. These 'constrain' it to behave in the finite element manner.

Still not a very interesting case? But it soon leads us back to the computer, and into the real world of engineering. Mathematically, it is exactly equivalent to the problem of heat-conduction, with w = temperature, so that the heat flows 'downhill' from hotter to colder regions:

(Heat flow per unit area, heat 'flux') =
(conductivity) × (temperature gradient).

Or it could be groundwater flow, with w = pore water pressure:

('flux' of water, per unit area) =
(permeability) × (pressure gradient).

Or it could be an electrical current, and by Ohm's law:

(current per unit area) = (conductivity) × (voltage gradient).

2.4 RIVERS OF HEAT

We are all interested in keeping warm, and Fig. 2.4 shows a crude model of heat flowing from a small square to a larger. We assume infinitely many 'cables', and the answers are not exactly the same as those given by a few discrete cables along the element boundaries, as in Fig. 2.3(a).

For FEMSKI, it should not matter how thick the equivalent membrane is, nor what the tensions are, provided that it is weightless, and isotropic (the same in all directions): for there are no external loads, only prescribed values along the boundaries. The input is summarized in the data file (Table 2.1). There is only one thing that you have not seen before: the prescribed value of 100°C at nodes 17 … 21.

The answers will obviously be bad near node 19 in Fig. 2.4. We return to the same problem in Fig. 2.5, this time with all guns blazing, and we are especially proud of the graphical presentation. Together, the flow-lines give an impressive picture of the overall flow, because their density varies as the local rate of flow. The 'flow-concentration' at the corner is very black – in fact it is an infinite concentration, although our calculations cannot show this.

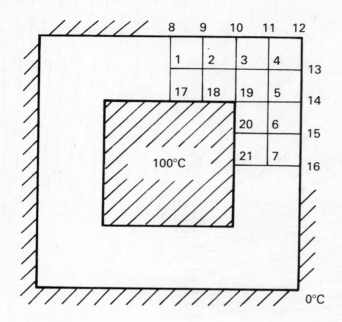

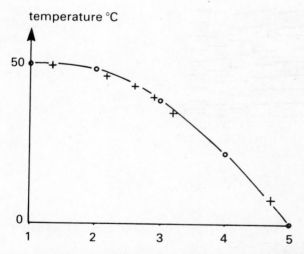

Fig. 2.4 – A steady-state heat-flow problem. By symmetry there are only four distinct nodal variables, the temperatures at the nodes 1 to 4, shown as dots on the graph. However, the crosses are temperatures estimated from the finer mesh of Fig. 2.5 and the agreement is quite good. See also Table 2.1.

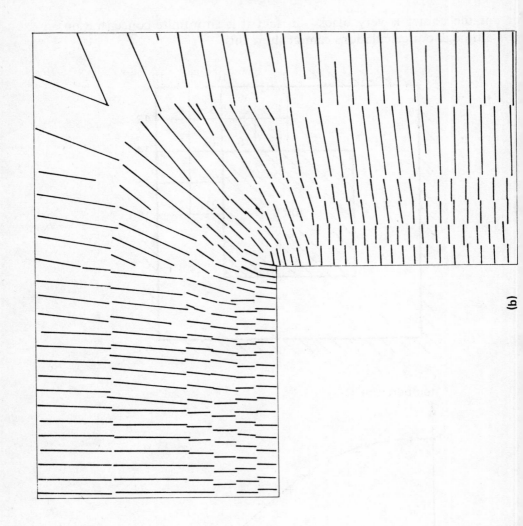

(b)

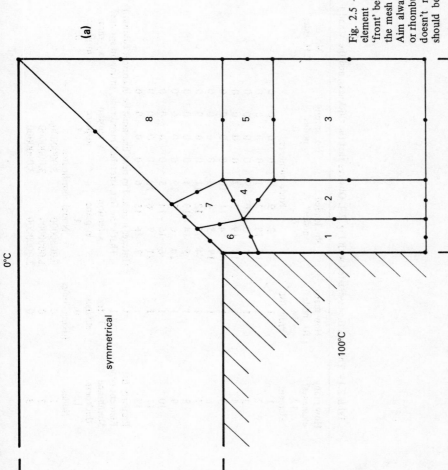

Fig. 2.5 – (a) A finer mesh, with 35 nodes. The element order (roman numerals) seeks a small 'front' between the mesh as far as element N, and the mesh to come. See definition of 'front solvers'. Aim always to have squares or squarish rectangles, or rhombuses: or maybe equilateral triangles, but it doesn't matter too much. The small elements should be even smaller. (b) Yet the flow-pattern is quite convincing.

Table 2.1 – The data file associated with Fig. 2.4. Observe that the thickness, and the x and y tensile stresses, are unity (arbitrary).

How many elements?	How many fix. nodes?	No. loaded nodes?	No. sprung nodes?	No. of R.H. sides?	Maximum R.H. sides?	Iterative solutions?
12	14	0	0	1	1	0

Element	Prop. list	Node numbers					
1	1	1	2	9	8	0	0
2	1	17	18	2	1	0	0
3	1	2	3	10	9	0	0
4	1	18	19	3	2	0	0
5	1	3	4	11	10	0	0
6	1	19	5	4	3	0	0
7	1	4	13	12	11	0	0
8	1	5	14	13	4	0	0
9	1	20	6	5	19	0	0
10	1	6	15	14	5	0	0
11	1	21	7	6	20	0	0
12	1	7	16	15	6	0	0

(Integer here and in next line must be in correct column.)

— That's the flat, stretched membrane, sitting on springs?

Property list

Element type: 1

Membrane thickness	x-tension in-plane	y-tension in-plane	xy-shear in-plane	Lateral pressure	Number of Gauss pnts.
1.	1.	1.	0.	0.	2.

Extra stiffness: 0.

Nodes	Fixing-codes	Nodal coordinates	
1	0	0.0000000	3.0000000
2	0	1.0000000	3.0000000
3	0	2.0000000	3.0000000

Node	Fixing-code		
4	0	3.0000000	3.0000000
5	0	3.0000000	2.0000000
6	0	3.0000000	1.0000000
7	0	0.0000000	0.0000000
8	1	0.0000000	4.0000000
9	1	1.0000000	4.0000000
10	1	2.0000000	4.0000000
11	1	3.0000000	4.0000000
12	1	4.0000000	4.0000000
13	1	4.0000000	3.0000000
14	1	4.0000000	2.0000000
15	1	4.0000000	1.0000000
16	1	4.0000000	0.0000000
17	0	0.0000000	2.0000000
18	0	1.0000000	2.0000000
19	0	2.0000000	2.0000000
20	0	2.0000000	1.0000000
-21	0	2.0000000	0.0000000

Node	Fixing-code	Nonzero prescribed values
17	1	100.000000
18	1	100.000000
19	1	100.000000
20	1	100.000000
21	1	100.000000

There are no extra loads, apart from gravity loads etc.
There are no additional springs to earth.

The most interesting property of flow-lines can be deduced by argument alone. These curves lie along the flow, everywhere. Thus there is no flow across any line. Therefore the total amount of heat/ water/electric current passing between any pair of lines remains strictly constant, from one end of a 'tube of flow' to the other, from source to mouth of this curving mini-river.

Alas, it will not be so in Chapter 6, when we consider the equivalent 'tube of stress'. This will represent a load; imagine a bunch of reinforcing fibres, each carrying its tension through the material, from region to region. But the force transmitted by a fibre is *not* constant. It can be influenced by the *other* fibres, the *other* stresses, even although they remain perpendicular; if they are curved, then we have a phenomenon that reminds us of 'hoop stress'. Remember, the tensions around the skin of a balloon support the pressure inside, although they are normal to the pressure loads everywhere.

2.5 SPRING SUPPORTED MEMBRANES

Back to structures; these will be our main discipline. Consider now a case in 2D that is simple enough to work by hand, and that carries the necessary teaching messages. Figure 2.6 is of no structural interest, but at least it looks something like a real finite element job. Also it embodies very clearly all the four general ideas. (So far, we have seen only one.)

It models very crudely a heavy, flexible membrane, supported on an elastic foundation. The triangles become rigid sheets joined by

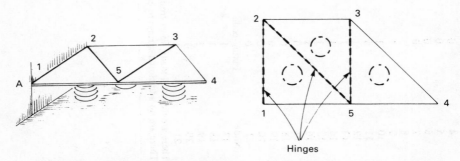

Fig. 2.6 – These three triangular membrane elements are hinged, either to each other or to the vertical support A. Each is supported by a spring at its centroid.

hinges: the physical form of the new shape function constraints is shown in Fig. 2.7. Compare these with the tubes around the cable, and the needles embedded in the previous membrane. This is another manifestation of our first general idea, *SHAPE FUNCTION CONSTRAINTS*.

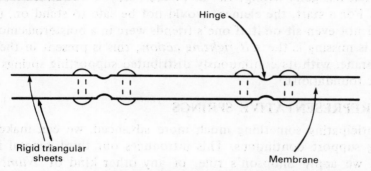

Fig. 2.7 – Each of the three elements is reinforced, by having rigid plates riveted to its top and bottom surface. These rigid plates force the out-of-plane deflections to vary linearly; hence the name, 'Shape Function Constraints'.

The elements in Fig. 2.6 connect at the hinges, all along their common edges. Of course, the computer only sees the values at the nodes, so from the computer's point of view the elements connect and interact only at the nodes. Basic to the whole finite element concept is the fact that the deflection at node 5 (Fig. 2.6) for example is common to all the three elements that meet there. This is the second general idea, *CONTINUITY AT NODES*. Very important.

Yes, very important ... Oops. See Fig. 2.8(a) for the answers – they are *horrible*. If the weight of each element is L and if the

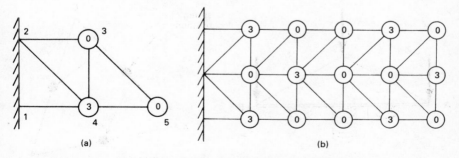

Fig. 2.8 – The ringed numbers show the deflections of the nodes. In (a) the results for the mesh in Fig. 2.5 are shown, with unit weight and unit spring stiffness per element: also in (b) the response of a finer mesh. This can, and will, be improved upon.

corresponding spring rate is k, then every centroid descends exactly $L/k = 1$, because this is a simple structure. But the correct deflection is 1 at every *node* — everywhere, in fact. At least all the *springs* are correctly loaded. Could be worse.

What has gone wrong? Let us ponder. Maybe we can learn something. For a start, the element would not be safe to stand on. (One would not even sit on it, if one's friends were in a boisterous mood.) What is missing is the *self-righting action*; this is present in the real membrane, with its continuously distributed supporting springs (the elastic foundation).

2.6 REPRESENTATIVE SPRINGS

By anticipating something much more advanced, we can make the spring support continuous. This introduces our third general idea. When we apply Simpson's rule, or any other kind of *'Numerical Integration'*, we sample the value of a function at certain representative points; we assume we know enough about how it behaves elsewhere. We take a few *'REPRESENTATIVE VALUES'* of depth, say, and we estimate roughly how much water there is in a pond. In the case of the spring supported triangle, we *replace* the continuous support of total rate k by a much simpler physical concept: a separate spring of rate $\frac{1}{3}k$ at each of the three midsides — see Fig. 2.9 — so that the integral (the strain energy) is nearly correct.

Let us imagine that the element vertices A, B, and C are deflected downwards, by three hydraulic jacks. The reactions R_A, R_B and R_C are measured by three load cells. Let us depress A a distance w_A, B a

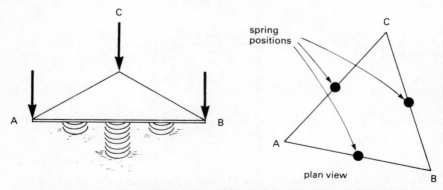

Fig. 2.9 — Better than a single spring at the centroid, is to have a number of them at carefully chosen positions. This choice gives the same answers as the exact method, which is much more difficult.

distance w_B, and C a distance w_C. Thus the spring midway along **AB** goes down the mean distance $\frac{1}{2}(w_A + w_B)$ and generates a force of $\frac{1}{6}k(w_A + w_B)$. This is the equivalent to a load of $\frac{1}{12}k(w_A + w_B)$ both at A and at B. In the same way, the spring midway along BC produces a load of $\frac{1}{12}k(w_B + w_C)$ both at B and at C. Finally we have a load of $\frac{1}{12}k(w_C + w_A)$ both at C and at A. Adding the effects of all three springs, we have a total load at A for example of $\frac{1}{12}k(2w_A + w_B + w_C)$ = R_A, the force registered by the load cell.

And so on. Hurrah for matrices. It all looks so easy:

$$\frac{1}{12}k \begin{bmatrix} 2 & 1 & 1 \\ 1 & 2 & 1 \\ 1 & 1 & 2 \end{bmatrix} \begin{bmatrix} w_A \\ w_B \\ w_C \end{bmatrix} = \begin{bmatrix} R_A \\ R_B \\ R_C \end{bmatrix}$$

This 3×3 is an element stiffness matrix. The *REPRESENTATIVE SPRINGS* that we used in finding it are our third general idea. These three equations are the ones to use for a one-element problem. However, real problems have many elements — many equations! The stiffness matrix 'K' becomes very large, but we still solve the basic equation

$$\mathbf{Kd} = \mathbf{F}$$

where   **K** = stiffness

   d = displacement

   F = force.

2.7 THE STRUCTURAL TEST-RIG

This brings us to our fourth and last general idea, the *ASSEMBLY OF ELEMENTS*. Imagine that we now have not three, but twelve jacks, one for each of the element vertices in Fig. 2.10. When only the first element is assembled into the big test-rig, we re-name the nodes A = 1, B = 2, C = 4:

$$\frac{1}{12}k \begin{bmatrix} 2 & 1 & 1 \\ 1 & 2 & 1 \\ 1 & 1 & 2 \end{bmatrix} \begin{bmatrix} w_1 \\ w_2 \\ w_4 \end{bmatrix} = \begin{bmatrix} R_1 \\ R_2 \\ R_4 \end{bmatrix} + \begin{bmatrix} \frac{1}{3}L \\ \frac{1}{3}L \\ \frac{1}{3}L \end{bmatrix}$$

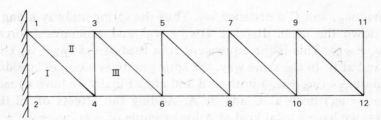

Fig. 2.10 – The lower half of Fig. 2.8(b) in plan view, invoking symmetry. This
time it is solved with continuous spring support.

Here we introduce the weight of the element, L, as well as the
reactions. So far, R_3 and R_5, R_6 ... are zero. Eventually we want
twelve consecutive equations, so we rearrange the coefficients of our
first element:

$$\frac{1}{12}k \begin{bmatrix} 2 & 1 & 0 & 1 \cdots \\ 1 & 2 & 0 & 1 \\ 0 & 0 & 0 & 0 \\ 1 & 1 & 0 & 2 \\ \vdots & & & \end{bmatrix} \begin{bmatrix} w_1 \\ w_2 \\ w_3 \\ w_4 \\ \vdots \end{bmatrix} = \begin{bmatrix} R_1 \\ R_2 \\ R_3 \\ R_4 \\ \vdots \end{bmatrix} + \frac{1}{3}L \begin{bmatrix} 1 \\ 1 \\ 0 \\ 1 \\ \vdots \end{bmatrix}$$

It is a question of housekeeping. We have two sets of addresses for
the coefficients. For the second element, A = 1, B = 3, C = 4:

$$\frac{1}{12}k \begin{bmatrix} 2 & 1 & 1 \\ 1 & 2 & 1 \\ 1 & 1 & 2 \end{bmatrix} \begin{bmatrix} w_1 \\ w_3 \\ w_4 \end{bmatrix} = \begin{bmatrix} R_1 \\ R_3 \\ R_4 \end{bmatrix} + \frac{1}{3}L \begin{bmatrix} 1 \\ 1 \\ 1 \end{bmatrix}$$

Taking the second element out of the element test-rig and putting it
with the first, into the larger rig, we have:

$$\frac{1}{12}k \begin{bmatrix} 4 & 1 & 1 & 2 \cdots \\ 1 & 2 & 0 & 1 \\ 1 & 0 & 2 & 1 \\ 2 & 1 & 1 & 4 \\ \vdots & & & \end{bmatrix} \begin{bmatrix} w_1 \\ w_2 \\ w_3 \\ w_4 \\ \vdots \end{bmatrix} = \begin{bmatrix} R_1 \\ R_2 \\ R_3 \\ R_4 \\ \vdots \end{bmatrix} + \frac{1}{3}L \begin{bmatrix} 2 \\ 1 \\ 1 \\ 2 \\ \vdots \end{bmatrix}$$

So the process continues. Fortunately, the computer revels in such boring tasks. Central to this task is the concept of the *destination vector*, a list of integers like $(1,2,4)$ for the first element, or $(1,3,4)$ for the second. The destination vector relates the physical addresses of the nodes in the element rig, to those in the larger rig. It expresses on paper, or in the computer, our fourth general idea, the *Assembly of Elements*.† To help your computer in this regard, determine if your program is a 'front' or 'band' solver, and follow the general aims indicated in our definitions.

Exercise: Copy the picture and the element matrix. Now reproduce the equations in Table 2.3 without cheating (too much). (You had better read on a little. The first two rows and columns are missing.)

2.8 SOLVING THE EQUATIONS

Having assembled the elements, we delete the first two equations and the first two columns, because $w_1 = w_2 = 0$. The other jacks can now be removed, giving $R_3 \ldots R_{12} = 0$. Each node is in equilibrium, free to move, except the first two. We then solve the equations with a pocket calculator, using a variant of Crout's tabular layout (see Table 2.2). It will be an hour or so well spent, to check our arithmetic in these examples.

But maybe you have never seen Crout's layout. Let us take, for example,

$$(\text{equation one}) \quad 4x + 3y + z = 14$$
$$(\text{equation two}) \quad 3x + 4y + 3z = 19$$
$$(\text{equation three}) \quad x + 3y + 4z = 17$$

The layout is shown in Table 2.2.

The 0.75 in the first column of the table is the number you would have to multiply 'equation one' by, before subtracting it from 'equation two', so as to eliminate x from 'equation two'. The other numbers in this row are the coefficients for y and z, and the right-hand-side in the reduced equation two, when 'equation one'

† To digress. Things are more complicated when each node has more than one variable, as in section 1.2. When the number of variables is different from node to node, as in the 'Semiloof' shell element, even a professional programmer will have to think carefully. To make things worse, we normally keep only a submatrix in core, because the full matrix is much too large. Housekeeping is a major headache in finite element programming.

is actually multiplied by 0.75 and subtracted from 'equation two'. For y, $4 - 3(0.75) = 1.75$. To check the work so far, we sum all the numbers to the right of the vertical line between columns 1 and 2: $1.75 + 2.25 + 8.5 = 12.5$.

Table 2.2

The tabular layout for the worked example. The check column is effectively another set of RHS (= right-hand sides) to be worked concurrently. For example, $22 = 4 + 3 + 1 + 14$, and $12.5 = 1.75 + 2.25 + 8.5$ — the sum of everything to the right of the wavy line. The final check row comprises the answers using the check RHS, giving $x' = x + 1$ etc.

The unknowns	x	y	z	RHS	Check RHS
The coefficients of the	4	3	1	14	22
unknowns on the equa-	3	4	3	19	29
tion left-hand sides.	1	3	4	17	25
Above the steps, the	4	3	1	14	22
coefficients of the	0.75	1.75	2.25	8.5	12.5
reduced equations.	0.25	1.286	0.857	2.571	3.429
Answer, x, y, z	2.0	1.0	3.0		
Check, x', y', z'	3.0	2.0	4.0		

In the next row down, the 0.25 is the number that 'equation one' has to be multiplied by, such that x is eliminated on subtraction from 'equation three'. This would leave $2.25 y + 3.75 z = 13.5$. The reduced equation two; $1.75 y + 2.25 z = 8.5$, is now multiplied by $9/7 = 1.286$ — watch for roundoff — and also subtracted from 'equation three', in order to eliminate y too. Thus the reduced coefficient for z in this row is $4 - (0.25)(1) - (1.286)(2.25) = 0.857$. We are now left with only one coefficient and a right-hand side: $0.857 z = 2.571$. We check the sum as before, and then calculate $z = 3.0$ (see the ANSWER row) — if we had kept enough figures in our calculator! Dividing the 'check' sum by the same coefficient gives $z = 4.0$. Because this is 1.0 more than 3.0, we have still not made any mistakes. Knowing $z = 3.0$ we can go to the 'reduced equation two' and solve for y, then check again: and so on.

With such regular checks, it is just about humanly possible to solve a large number of equations in a reasonable time. Before computers entered the engineering scene, many people owed their sanity to methods like this one. Who knows, one day *you* might need it ....

Now, let us return to the real problem, in Fig. 2.10 and Table 2.3. The answers are fascinating, to compare with those for the element with only one spring. Quite a contrast. The commotion near the support is now damped, very rapidly. There is hardly any noise in the last two elements. The solitary rogue value, $w_2 = 1.51$, corresponds to a value of 3 in Fig. 2.8(b).

In conclusion, we emphasize that the problem is very tricky indeed. At the support, $w = 0$, and immediately to the right, $w = 1$ in the analytical solution. We have succeeded remarkably well, considering the sudden jump in w.

The exposition of finite elements is also proceeding well, and some of the main points in the technique are becoming clear. Finite elements are models involving shape function constraints, which make them connect not only at nodes but also along every common edge. The distributed springiness of the material can be 'lumped' into a few representative springs. When the element stiffness matrices become available, they must be assembled, like the elements themselves.

These principles need not be submerged under a mass of complicated details, introductory examples must be simple. Our examples admittedly bear little resemblance to a car body. But even the specialists in crash simulation still have difficulties in modelling on their computers the tangled wrecks that litter our highways.

2.9 HI FEMSKI! – THE HANGING SIGNPOST

A little more computing. In Chapter 1 we stressed a plate loaded in its own plane – see Fig. 1.1. This type of exercise is known as 'plane stress'. Let us do a second computer run; this time there will be a glimmering of understanding. In the membrane triangle we make w vary linearly within each element. We now apply the same idea to plane stress. We make *both* u and v, the deflections in the x- and y-directions, vary linearly with x and y.

If we had been interested, we could have found the slopes $\partial w/\partial x$ and $\partial w/\partial y$ in the membrane element. They were constant. Likewise,

Table 2.3

The assembled equations for Fig. 2.10 are solved. The conditioning is very exceptional: usually at least 8 decimals are desirable for finite elements. † is the sum of the coefficients and RHS. Equations with these RHS would have their roots increased by 1. * means that 0.061 times equation 2 (reduced) is to be subtracted from equation 4. # means that, allowing for roundoff, there are no errors so far, because 2.005 is sufficiently close to 1.007 + 1.

Assembled equations

w_3	w_4	w_5	w_6	w_7	w_8	w_9	w_{10}	w_{11}	w_{12}	RHS	Check sum
0.500	0.167	0.083	0.167	0	0	0	0	0	0	1.000	1.917†
0.167	0.500	0	0.083	0	0	0	0	0	0	1.000	1.750
0.083	0	0.500	0.167	0.083	0.167	0	0	0	0	1.000	2.000
0.167	0.083	0.167	0.500	0	0.083	0	0	0	0	1.000	2.000
0	0	0.083	0	0.500	0.167	0.083	0.167	0	0	1.000	2.000
0	0	0.167	0.083	0.167	0.500	0.083	0	0	0	1.000	2.000
0	0	0	0	0.083	0.083	0.500	0.167	0	0.167	1.000	2.000
0	0	0	0	0.167	0	0.167	0.500	0.083	0.083	1.000	2.000
0	0	0	0	0	0	0	0.083	0.167	0.083	0.333	0.666
0	0	0	0	0	0	0.167	0.083	0.083	0.333	0.667	1.333

REDUCED COEFFICIENTS

w_3	w_4	w_5	w_6	w_7	w_8	w_9	w_{10}	w_{11}	w_{12}	RHS	Check sum
0.500	0.167	0.083	0.167							1.000	1.917
	0.444	−0.028	0.027							0.666	1.110
		0.484	0.141	0.083	0.167					0.875	1.750
			0.402	−0.024	0.034					0.371	0.783
				0.484	0.140	0.083	0.167			0.873	1.748
					0.399	−0.024	0.035			0.413	0.822
						0.484	0.141	0.083	0.167	0.875	1.750
							0.398	−0.024	0.035	0.409	0.818
								0.151	0.057	0.208	0.416
									0.251	0.252	0.503

MULTIPLIERS USED DURING REDUCTION

Column	Multipliers
w_3	0.334, 0.166, 0.334
w_4	−0.062, 0.061*
w_5	0.291, 0.171, 0.345
w_6	−0.060, 0.086
w_7	0.290, 0.171, 0.345
w_8	−0.060, 0.087
w_9	0.290, 0.171, 0.345
w_{10}	−0.060, 0.087
w_{11}	0.374

ANSWERS = w_i CHECK = w_i + 1

w_3	w_4	w_5	w_6	w_7	w_8	w_9	w_{10}	w_{11}	w_{12}
1.025	1.510	1.028	0.897	0.996	1.007	0.999	1.000	0.998	1.004
(2.024)	(2.512)	(2.028)	(1.898)	(1.999)	(2.005)#	(1.999)	(2.000)	(1.998)	(2.004)

the strains here, $\partial u / \partial x = \epsilon_{xx}$ etc. and hence also the stresses σ_{xx} etc. are constant over each element. Whenever the deflections u and v vary linearly with x and y, the strains are constant. This is why people often speak of the 'constant stress triangle'. If these remarks are meaningless, please turn to Chapter 6. Using techniques we shall begin to explain in the next chapter, it is possible to find the stiffness matrix. In fact, this is the simplest 2D element. (It does not work very well.)

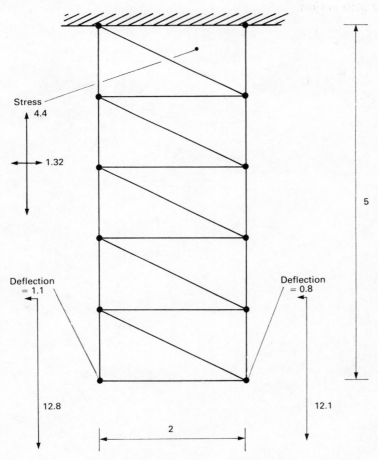

Fig. 2.11 – Ten linear triangles represent a vertical column in plane stress, loaded by gravity. The vertical stress as shown should be $4\frac{2}{3}$. The horizontal tensile stress is genuine, because the fixing prevents the contraction due to Poisson's ratio = 0.3. The bottom edge should descend $12\frac{1}{2}$ uniformly. The movements to the right, and the rotation, are completely spurious: see text.

The assignment in Fig. 2.11, or something like it, is little more than a practice run. Nevertheless, the results are encouraging. Of course, the spurious sideways motions are small, and we must not be too purist. The right-hand triangles bear slightly more gravity load than the left-hand triangles. This sort of thing can always happen when a mesh is not symmetrical, is not a mirror-image of itself. But if the problem is symmetrical and the answers are *not,* then you have evidence of Satanic intervention, and the trouble-shooter in you must jump into action.

CHAPTER 3

What do we actually do?

3.1 WORK TO BE DONE

Train yourself always to think *strain energy*. If the energy is right, not only for what actually happens, but for slightly different responses too, then the theory has to be correct. This idea of 'virtual work' is very powerful.† And when the strain energy is written down in a special way, the stiffness matrix drops out, as if by magic. Remember, that is what we want — the stiffness matrix.

We cannot avoid ordinary integration at some stage, to find the strain energy. This is drudgery; in the end, however, it leads us back to the lumped-spring concept, and things are simple again. The previous chapter used lumped-springs too, but in an intuitive way. We need more science in real elements.

3.2 THE BILINEAR SQUARE

The element shown in Fig. 3.1(a) is the 2 × 2 square in Fig. 3.2(a). The total stiffness $4k$ is spread uniformly, so that the strain energy is generated by an infinity of small springs of rate $k \, dx \, dy$. The total weight is L, so that $\frac{1}{4}L$ must be supported at each corner. The formula for the 'shape function' in Fig. 3.1(a) is

$$N_C = \tfrac{1}{4}(1 + x)(1 + y)$$

because this is 1 at C, and zero at A, B and D. Similarly:

$$N_A = \tfrac{1}{4}(1 - x)(1 - y)$$
$$N_B = \tfrac{1}{4}(1 + x)(1 - y)$$
$$N_D = \tfrac{1}{4}(1 - x)(1 + y)$$

† 'Reckon thine energy faithfully, which shall neither increase nor diminish, and shall be conserved for all thy days.' (Eleventh commandment, 'Authorized' version!)

by replacing x by $-x$ and/or y by $-y$. Our purpose here is to write the deflection at any point in the element in the simple form

$$w = N_A w_A + N_B w_B + N_C w_C + N_D w_D \qquad (3.1)$$

without bending any of the constraint needles. See Fig. 3.1(b). Thus, if we know the deflections of the four corners, we know the deflection of any point. You can see now why the N are zero or 1. In this course we create the formulae for shape functions by trial and error, plus a little intelligence.

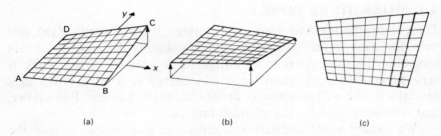

(a)   (b)   (c)

Fig. 3.1 – (a) The shape function N_C which is unity at C, zero at B, D, and A. (b) The shape function constraints are rigid needles in the surface. (c) The Taig quadrilateral, in plan view. The needles intersect the sides at uniform intervals.

3.3 THE USE OF STRAIN ENERGY

Let us recall the formula for the strain energy in a spring, $U = \frac{1}{2}kw^2$ where k is the rate of the spring.† Here we have k per unit area, giving $k\,dx\,dy$ as the rate for the infinitesimal spring associated with infinitesimal area $dx\,dy$ for the square of Fig. 3.2(a). Integrating over the whole area gives the total strain energy

$$U = \int_{-1}^{1} \int_{-1}^{1} \tfrac{1}{2}k\,w^2 \,dx\,dy.$$

† For a spring, $R = kw$, and the strain energy is the total work done in compressing it:

$$\int R\,dw = \int kw\,dw = [\tfrac{1}{2}kw^2]_{w=0}^{w=w\,\text{final}}$$
$$= \tfrac{1}{2}kw^2_{\text{final}}$$
$$= \tfrac{1}{2}R_{\text{final}}\,w_{\text{final}}.$$

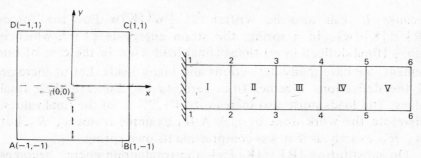

Fig. 3.2 – (a) The coordinate system for the square element, in order to write the
shape functions. (b) The membrane problem to be solved (looking downwards –
this is not a beam problem).

Substituting for w,

$$
w = [w_A \, w_B \, w_C \, w_D] \begin{bmatrix} N_A \\ N_B \\ N_C \\ N_D \end{bmatrix} \quad \text{or} \quad [N_A \ N_B \ N_C \ N_D] \begin{bmatrix} w_A \\ w_B \\ w_C \\ w_D \end{bmatrix}
$$

we can write the strain energy in its 'palindromic' form:‡

$$
U = \tfrac{1}{2} [w_A \ w_B \ w_C \ w_D] \int_{-1}^{1} \int_{-1}^{1} \begin{bmatrix} N_A \\ N_B \\ N_C \\ N_D \end{bmatrix} k \, [N_A \ N_B \ N_C \ N_D] \, \mathrm{d}x \, \mathrm{d}y \begin{bmatrix} w_A \\ w_B \\ w_C \\ w_D \end{bmatrix}.
$$

We are allowed to remove $w_A \ldots w_D$ from the integrand, because
these quantities do not vary with x and y. What is left inside the box
is the *element stiffness matrix*:

$$
[K] = \begin{bmatrix} K_{AA} & K_{AB} & \cdots \\ K_{BA} & & \\ \vdots & & \end{bmatrix}
$$

‡ The same backwards, like the word RADAR, ignoring $\tfrac{1}{2}$, $\int\!\int$ and $\mathrm{d}x \, \mathrm{d}y$.

because U can also be written as $\frac{1}{2}\mathbf{w}^T[\mathbf{K}]\mathbf{w}$. For the forces $\{R\}=[\mathbf{K}]\{\mathbf{w}\}$. In a spring, the strain energy is $\frac{1}{2}kw^2$, which is also $\frac{1}{2}$ (final deflection w) times (final load kw). In the case of our element, we have many deflections and many loads. Let us increase all the deflections together from zero, to 1%, 2% ... of their final values. The loads must also increase to 1%, 2% ... of the final values. Therefore the work done by jack A for example is not $w_A R_A$ but $\frac{1}{2}w_A R_A$, exactly as if it was compressing its own spring.

On substituting $\{R\}=[\mathbf{K}]\{\mathbf{w}\}$, the total strain energy becomes $\frac{1}{2}(w_A R_A + w_B R_B + w_C R_C + w_D R_D)=\frac{1}{2}\{\mathbf{w}\}^T[\mathbf{K}]\{\mathbf{w}\}$ as required. This shows that $[\mathbf{K}]$ is indeed the matrix inside the box. It now remains only to compute $[\mathbf{K}]$:

$$k\int_{-1}^{1}\int_{-1}^{1}\begin{bmatrix}(1-x)^2(1-y)^2, (1-x^2)(1-y)^2, (1-x^2)(1-y^2), (1-x)^2(1-y^2)\\ (1-x^2)(1-y)^2, (1+x)^2(1-y)^2, (1+x)^2(1-y^2), (1-x^2)(1-y^2)\\ (1-x^2)(1-y^2), (1+x)^2(1-y^2), (1+x)^2(1+y)^2, (1-x^2)(1+y)^2\\ (1-x)^2(1-y^2), (1-x^2)(1-y^2), (1-x^2)(1+y)^2, (1-x)^2(1+y)^2\end{bmatrix}dx\,dy$$

$$=\frac{k}{9}\begin{bmatrix}4&2&1&2\\2&4&2&1\\1&2&4&2\\2&1&2&4\end{bmatrix}$$

Let us use $\mathbf{K}$, to solve the problem in Fig. 3.2(b). We assume that the deflections at the two sides are identical; this saves arithmetic. Thus the destination vector for element I in the mesh of Fig. 3.2(b) is [1, 2, 2, 1], because for example w_2 would depress B and C simultaneously. Assembling, and deleting the first row and column as before, we have Table 3.1. The answers are very similar in Table 2.2, except that we have lost the worst rogue value. This is a little disappointing. But for the moment, we are trying to develop new techniques, rather than to improve the answers.

3.4 THE TAIG QUADRILATERAL

Casting our minds back to the crumpled wreck of a car, it is evidently progress if we can use elements of various shapes. The simplifying feature of the Taig quadrilateral is that it uses *local* coordinates ξ and

η (xi and eta). In Fig. 3.2(a), x and y go from -1 to 1. In Fig. 3.3, ξ and η also go from -1 to 1; ξ replaces x, and η replaces y. In this sense it is still a 'square'.

Table 3.1 – The solution of the problem in Fig. 3.2(b) using the bilinear membrane element, with $k = \frac{1}{4}$.

w_2	w_3	w_4	w_5	w_6	RHS	Check sum
0.667	0.167	0	0	0	1	1.834
0.167	0.667	0.167	0	0	1	2.001
0	0.167	0.667	0.167	0	1	2.001
0	0	0.167	0.667	0.167	1	2.001
0	0	0	0.167	0.333	0.5	1
0.667	0.167	REDUCED			1	1.834
0.250	0.625	0.167	COEFFICIENTS		0.750	1.543
	0.267	0.622	0.167		0.800	1.589
MULTIPLIERS USED		0.268	0.622	0.167	0.788	1.575
DURING REDUCTION			0.268	0.288	0.289	0.578
1.267	0.928	1.019	0.994	1.003	ANSWERS	
2.267	1.929	2.020	1.993	2.007	CHECK	

The theory is outlined in Fig. 3.3. Let us walk to point R, 63% of the distance from A to B. Because 50% would give $\xi = 0$, halfway, we have $\xi = 0.26$ for our 63%. Thus

$$\begin{bmatrix} x_R \\ y_R \end{bmatrix} = 0.37 \begin{bmatrix} x_A \\ y_A \end{bmatrix} + 0.63 \begin{bmatrix} x_B \\ y_B \end{bmatrix} = \frac{1-\xi}{2} \begin{bmatrix} x_A \\ y_A \end{bmatrix} + \frac{1+\xi}{2} \begin{bmatrix} x_B \\ y_B \end{bmatrix}$$

Walking similarly along DC, to S:

$$\begin{bmatrix} x_S \\ y_S \end{bmatrix} = \frac{1-\xi}{2} \begin{bmatrix} x_D \\ y_D \end{bmatrix} + \frac{1+\xi}{2} \begin{bmatrix} x_C \\ y_C \end{bmatrix}$$

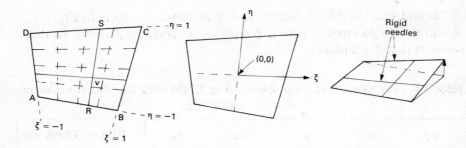

Fig. 3.3 – The Taig quadrilateral. The distance, measured along any line ξ = constant, varies linearly with η, and vice versa. Every line ξ = constant or η = constant is a rigid needle.

Let us now take a walk up RS:

$$\begin{bmatrix} x_V \\ y_V \end{bmatrix} = \frac{1-\eta}{2} \begin{bmatrix} x_R \\ y_R \end{bmatrix} + \frac{1+\eta}{2} \begin{bmatrix} x_S \\ y_S \end{bmatrix}$$

$$= \frac{1-\eta}{2} \left(\frac{1-\xi}{2} \begin{bmatrix} x_A \\ y_A \end{bmatrix} + \frac{1+\xi}{2} \begin{bmatrix} x_B \\ y_B \end{bmatrix} \right) +$$

$$\frac{1+\eta}{2} \left(\frac{1-\xi}{2} \begin{bmatrix} x_D \\ y_D \end{bmatrix} + \frac{1+\xi}{2} \begin{bmatrix} x_C \\ y_C \end{bmatrix} \right)$$

$$= N_A \begin{bmatrix} x_A \\ y_A \end{bmatrix} + N_B \begin{bmatrix} x_B \\ y_B \end{bmatrix} + N_C \begin{bmatrix} x_C \\ y_C \end{bmatrix} + N_D \begin{bmatrix} x_D \\ y_D \end{bmatrix}$$

where $N_A = \frac{1}{4}(1-\xi)(1-\eta)$
$N_B = \frac{1}{4}(1+\xi)(1-\eta)$
$N_C = \frac{1}{4}(1+\xi)(1+\eta)$
$N_D = \frac{1}{4}(1-\xi)(1+\eta).$

It is tempting to observe that $N_A \ldots N_D$ are identical in form to those of section 3.2. Can we use these too, as shape functions? Consider what would happen:

$$w = N_A w_A + N_B w_B + N_C w_C + N_D w_D.$$

At A, $\xi = \eta = -1$, so $N_A = 1$ and N_B, N_C and N_D are zero. Therefore $w = w_A$ as it should. And so on. Moreover, w varies linearly with ξ. But the distance along any line, $\eta =$ constant, also varies linearly with ξ. Therefore the needles in that direction do not bend. Nor do they in the η direction. So we have arrived. The N are indeed the shape functions. Remarkable? We still think so.

3.5 LUMPED SPRINGS FOR INTEGRATION – TAIG QUADRILATERAL

As in the square, we now want to integrate over the area of the quadrilateral to get

$$U = \tfrac{1}{2} \int k \, w^2 \, d(\text{area}).$$

The troublesome term will be d(area). We should prefer

$$U = \tfrac{1}{2} \int_{-1}^{1} \int_{-1}^{1} \phi \, d\xi \, d\eta$$

where ϕ is some new function of ξ and η. If d(area) = (scaling factor) $d\xi \, d\eta$, then our immediate problem is to find the scaling factor. See Fig. 3.4.

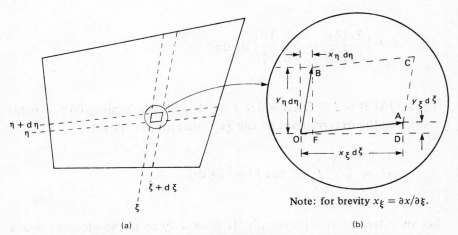

(a)   (b)

Note: for brevity $x_\xi = \partial x/\partial \xi$.

Fig. 3.4 – How to integrate over the Taig quadrilateral. The infinitesimal area $d\xi \, d\eta$ blows up into the parallelogram shown in (b): the smaller we make it, the more nearly parallel the sides become.

Let us first select a point (ξ, η). Next, we investigate the 'differential geometry' near that point. If we alter ξ and η by small amounts, what happens to x and y? This information will come from

$$
\mathbf{J} = \begin{bmatrix} \partial x/\partial \xi & \partial y/\partial \xi \\ \partial x/\partial \eta & \partial y/\partial \eta \end{bmatrix}
$$

$$
= \begin{bmatrix} \partial N_A/\partial \xi & \partial N_B/\partial \xi & \partial N_C/\partial \xi & \partial N_D/\partial \xi \\ \partial N_A/\partial \eta & \partial N_B/\partial \eta & \partial N_C/\partial \eta & \partial N_D/\partial \eta \end{bmatrix} \begin{bmatrix} x_A & y_A \\ \vdots & \vdots \\ x_D & y_D \end{bmatrix}
$$

$$
= \begin{bmatrix} -\frac{1}{4}(1-\eta), \frac{1}{4}(1-\eta), \frac{1}{4}(1+\eta), -\frac{1}{4}(1+\eta) \\ -\frac{1}{4}(1-\xi), -\frac{1}{4}(1+\xi), \frac{1}{4}(1+\xi), \frac{1}{4}(1-\xi) \end{bmatrix} \begin{bmatrix} x_A & y_A \\ \vdots & \vdots \\ x_D & y_D \end{bmatrix}
$$

$\mathbf{J}$ stands for 'Jacobian', a useful matrix. Here, for example, it defines the small area in question, in Fig. 3.4. Let us consider *half* the area of this little parallelogram, the triangle

$$\mathrm{OAB} = \mathrm{FBAD} + \mathrm{OBF} - \mathrm{ODA}$$

$$
= \frac{\left(\dfrac{\partial x}{\partial \xi}\, \mathrm{d}\xi - \dfrac{\partial x}{\partial \eta}\, \mathrm{d}\eta \right) \left(\dfrac{\partial y}{\partial \xi}\, \mathrm{d}\xi + \dfrac{\partial y}{\partial \eta}\, \mathrm{d}\eta \right)}{2} + \frac{\dfrac{\partial x}{\partial \eta} \dfrac{\partial y}{\partial \eta}\, \mathrm{d}\eta^2 - \dfrac{\partial x}{\partial \xi} \dfrac{\partial y}{\partial \xi}\, \mathrm{d}\xi^2}{2}
$$

$$
= \frac{1}{2}\left(\frac{\partial x}{\partial \xi} \frac{\partial y}{\partial \eta} - \frac{\partial x}{\partial \eta} \frac{\partial y}{\partial \xi} \right) \mathrm{d}\xi\, \mathrm{d}\eta .
$$

Thus, $\mathrm{OACB} = 2 \times \mathrm{OAB} = \det \mathbf{J}\, \mathrm{d}\xi\, \mathrm{d}\eta$. So the 'scaling factor' turns out to be the determinant of the Jacobian matrix, giving

$$
U = \frac{1}{2} \int\limits_{-1}^{1} \int\limits_{-1}^{1} (k \det \mathbf{J})\, w^2\, \mathrm{d}\xi\, \mathrm{d}\eta .
$$

Let us interpret this physically. It is exactly as if the element was a bilinear *square*, 2×2 in ξ and η, with *variable* support stiffness, $k^* = k \det \mathbf{J}$. The strain energy is the integral of $\phi = \frac{1}{2} k^* w^2$, over the

2×2 square. We use not the Simpson rule, but the Gauss 2-point rule, which is better in every way:

$$\int_{-1}^{1} \phi(z)\, dz = \phi(-\sqrt{\tfrac{1}{3}}) + \phi(\sqrt{\tfrac{1}{3}}) \tag{3.2}$$

$$= \phi(-a) + \phi(a), \text{ where } a = \sqrt{\tfrac{1}{3}}.$$

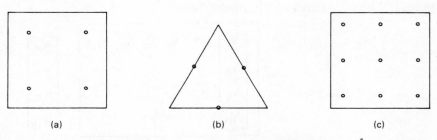

(a)   (b)   (c)

Fig. 3.5 – Some Gauss sampling points, for numerical integration, (a) $(\pm\sqrt{\tfrac{1}{3}} \pm \sqrt{\tfrac{1}{3}})$ for the Taig quadrilateral, (b) the midsides for a triangle, and (c) the 3×3 rule for a square. All of these are optimum choices, in different senses. FEMSKI allows you to choose.

Let us integrate first in ξ, keeping η constant:

$$\int_{-1}^{1} \left[\int_{-1}^{1} \phi(\xi,\eta)\, d\xi \right] d\eta = \int_{-1}^{1} [\phi(-a,\eta) + \phi(a,\eta)]\, d\eta$$

and then in η:

$$= [\phi(-a, -a) + \phi(-a, a)] + [\phi(a, -a) + \phi(a, a)].$$

This is known as a 'Gauss product rule', or simply as the Gauss 2×2. Fig. 3.5(a) shows the four Gauss points. Let us again interpret this physically. At each Gauss point in the square, we have a value of w, and of k^*. The strain energy is the *sum* of the four products, $\tfrac{1}{2} k^* w^2$. Therefore it is *exactly* as if we had four springs supporting the square, of *different* rates $k^* = k \det \mathbf{J}$.

To complete the physical picture, let us revert to the quadrilateral with the needles, and place the four lumped springs under the *mapped* Gauss points. Remember the triangle. But now the springs are stiffer where the quadrilateral bulges — which makes good intuitive sense.

3.6 THE STIFFNESS MATRIX – TAIG QUADRILATERAL

So we come full circle. Having involved ourselves with some tedious integration, we discover in the end that even the Taig quadrilateral has a simple physical message!

Should we continue in algebraic terms, or should we make the last step from the new physical model, with lumped springs? It makes no difference! The strain energy due to the spring at the first Gauss point is $\frac{1}{2}k^*w^2$, again in palindromic form:

$$\frac{1}{2}\begin{bmatrix} w_A & w_B & w_C & w_D \end{bmatrix} \begin{bmatrix} N_A \\ N_B \\ N_C \\ N_D \end{bmatrix} k^* \begin{bmatrix} N_A & N_B & N_C & N_D \end{bmatrix} \begin{bmatrix} w_A \\ w_B \\ w_C \\ w_D \end{bmatrix}$$

where $k^* = k \det \mathbf{J}$ is calculated for $\xi = \eta = -\sqrt{\frac{1}{3}}$ as are $N_A \ldots N_D$. The contribution to $\mathbf{K}$ is the 4×4 matrix within the frame. We accumulate to this the numbers that come from the second, third, and fourth Gauss points (Fig. 3.5) and the task is complete.

We have now learnt to use strain energy constructively, and we have found a very direct way to evaluate $[\mathbf{K}]$, which is gratifying. We have encountered a slightly less restrictive shape function constraint, which led us to a more general bilinear element, one that requires numerical integration.† Making a virtue out of necessity, we can now restore the most attractive feature of Chapter 2, the equivalent lumped springs at the Gauss points.

3.7 IN-PLANE DEFLECTIONS – TAIG QUADRILATERAL

Thus we are discussing what is called an 'isoparametric element' in some technical detail, near the end of the third chapter.‡ Finite elements are evidently less difficult than some people would like us

---

† This is approximate for the general quadrilateral, but it is always good enough to satisfy the 'patch test' for convergence.

‡ A silly name, really. No PARAMETERS are the same. The SHAPE FUNCTIONS are the same:

$$\begin{Bmatrix} x \\ y \end{Bmatrix} = \Sigma N_i \begin{Bmatrix} x_i \\ y_i \end{Bmatrix} \text{ and } \begin{Bmatrix} u \\ v \end{Bmatrix} = \Sigma N_i \begin{Bmatrix} u_i \\ v_i \end{Bmatrix}$$

to think. Admittedly we have restricted our attention to a membrane problem. But our aim is to present simple cases, which can be solved *by hand*. This is the only reason why our problems so far have been unrealistic.

We have seen problems which assume that the deflections are in the plane of the paper. Conceptually this is within our grasp; the new technical problems in applying the Taig quadrilateral are not unduly difficult either. The shape function constraints are now visualized as a large number of pantographs, as shown in Fig. 3.6, instead of the needles of Fig. 3.1. (The pantograph insists that u and v *both* vary linearly along its length.) We can nearly solve this class of problem. In fact, we are progressing very rapidly.

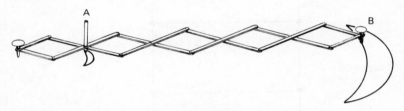

This is a pantograph

Fig. 3.6 — The pantograph was once widely used. The pencil at A draws a picture similar to that at B, but scaled down five times.

Using FEMSKI, we *can* already do this type of job with the Taig quadrilateral. Let us go to the computer, and solve the hanging signpost problem again. This is not a worthwhile assignment, but it is interesting to observe that Fig. 3.7 has only five elements, whereas Fig. 2.11 has ten. With any very fine, very extensive mesh there are nearly twice as many triangles as quadrilaterals, using the same nodes. The proof is devastatingly simple.† Quadrilaterals are a labour-saving device!

3.8 RESPECTABILITY: THE PATCH TEST FOR CONVERGENCE

It is always sound strategy to do a patch test, even if one is assured on the highest authority that a program is free of bugs. One asks

---

† The internal angles in an element add up to ... ? The total angle per internal node is 360°. Put N_n = number of nodes, and N_e = number of elements. What is the total angle in both cases?

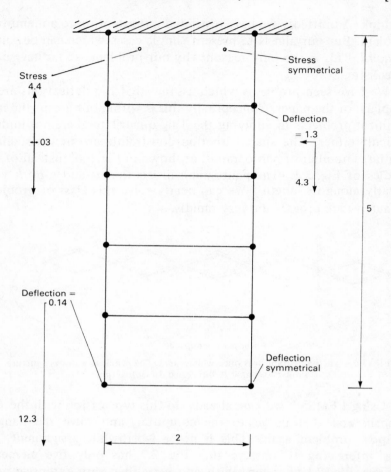

Fig. 3.7 – The same problem as in Fig. 2.11, but solved with five bilinear rectangles. It is not much better, except that the spurious sideways movements have gone. Observe the 'necking' due to Poisson's ratio, just below the support, and the shear which makes this possible is reflected in the slight rotation of the principal stresses.

the computer this question: can we define a state of uniform stress over a patch covered by *several* elements, if they are of arbitrary geometry? Figure 3.8(a), for example. This sounds slightly dotty. Suppose we had a complicated algorithm, to put springs in series as in Fig. 3.8(b). Would it make sense, if the load was *not* the same in all the springs? We wish that you could see a researcher sobbing gently, because a new element willingly accepts a constant stress, whatever the geometry, but a *patch* of the same elements persistently fails! It happens all too often.

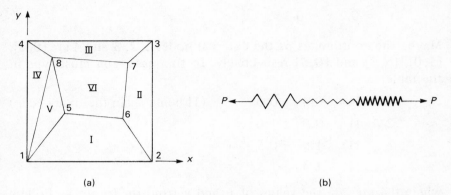

(a)   (b)

Fig. 3.8 – (a) The mesh chosen for the patch test. We should mix triangles and quadrilaterals in a patch like this, if we intend to mix them later in a real job. Nodes 1–4 are 'external'. We give them deflections according to some arbitrary state of constant strain. The 'internal' nodes 5–8 are unloaded, and we study their behaviour. (b) shows the equivalent test in one dimension.

If the test *succeeds* for arbitrary geometry, then we have an important research conclusion. Consider some stressing exercise on the car body, with very fine mesh. We expect good answers. Let us take a microscope to the real thing. We see a very small region. If it is small enough, the stresses will be sensibly constant. Thus we have a right to expect that the *finite element* version of the same region should give almost constant stress too. If a patch cannot do this, then we must give up any hope of convergence with an arbitrary fine mesh!

Conversely, mathematicians have now proved that if an element *passes* the patch test, then we have fine-mesh convergence, provided that the elements are not 'floppy' as in section 2.5. This is the finite element hallmark of excellence.

The patch test on Fig. 3.8(a), or something like it, is an assignment you must do. We insist. The data will be similar to those in section 1.3, except that when he asks you:

"Are there any NONZERO prescribed values?"

you have something to interest him, this time. Let us assume that the 'target state' for your patch test will be described by:

$$u = 0.1\,x + 0.2\,y + 0.4$$
$$v = 0.2\,x - 0.3\,y.$$

Maybe the coordinates of the external nodes 1, 2, 3 and 4 are [0, 0], [5, 0], [5, 5] and [0, 5] respectively. In that case, you must type in the table:

1	11	0.4	0.0	(11 being again the fixing code)
2	11	0.9	1.0	
3	11	1.9	−0.5	
4	11	1.4	−1.5	

where 0.4 etc. are the values of u and v from the formulae. Follow this by something naughty to violate the format, as before, like an isolated 'minus' sign.

When the job runs, you should find that all the stresses are equal, and that the deflections u and v at the other nodes also obey the formulae. Success! Many experienced finite element people have never *themselves* performed this crucial test. *You* have. Shoot that trouble before it arises! Devils beware!

CHAPTER 4

What do good elements do?

4.1 WHAT IS 'GOOD'?

All good elements should pass the patch test. But an element may pass the patch test without being 'good'. Explanations, later. For the moment, let us try a more elaborate element, again applied to the spring-supported membrane. With some patience, we can still use a pocket calculator throughout.

4.2 THE 8-NODE SQUARE ELEMENT

The element in Fig. 4.1 will teach us a lot, so it is worth the effort. Besides, it is a good performer, and is widely used. And being an isoparametric element, integrated numerically, it is easy to code, compared with many.

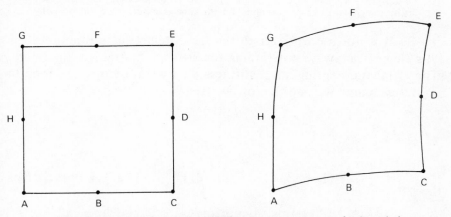

Fig 4.1 – With extra nodes at the midsides, we can assume quadratic variations in *w* within the element. This is rather like fitting a given curve with a series of parabolas, instead of straight lines. Indeed, the answers are much better. Furthermore, we can now have curved edges. This makes it easier to fit circles, etc.

We do not have time to philosophize over the shape functions, nor indeed the technical equipment. A typical midside shape function is

$$N_F = \tfrac{1}{2}(1 - x^2)(1 + y)$$

It has parabolas along x, with needles along y — for the moment. Part of Fig. 4.2 shows $-\tfrac{1}{2}N_F$.

The equation expressed pictorially in Fig. 4.2 is

$$N_E = \tfrac{1}{4}(1 + x)(1 + y) - \tfrac{1}{2}N_F - \tfrac{1}{2}N_D$$
$$= \tfrac{1}{4}(1 + x)(1 + y)(x + y - 1).$$

Reflecting to find the others means changing the sign of x or y, or interchanging them, very much as before. Observe carefully that N_E for example is zero at the other seven nodes. So we can write an equation like (3.1) but with eight terms.

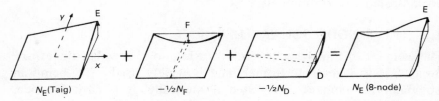

N_E(Taig) $-\tfrac{1}{2}N_F$ $-\tfrac{1}{2}N_D$ N_E (8-node)

Fig. 4.2 – To find N_E for the square element, we take the bilinear version and we subtract $\tfrac{1}{2}N_D$ and $\tfrac{1}{2}N_F$. No needles remain. To some extent, these diagrams justify the choice of shape functions, for the bilinear responses are still available.

So it is not too difficult to find acceptable shape functions, for this element anyway. But finding the stiffness matrix is a nightmare. The elementary spring has a stiffness $k\,dx\,dy$ as before. The term in **K** in the column w_E and the row w_F is

$$K_{EF} = k \int_{-1}^{1}\int_{-1}^{1} N_E N_F \, dx\, dy$$

$$= \frac{k}{8} \int_{-1}^{1}\int_{-1}^{1} (1 - x^2)(1 + y)(1 + x)(1 + y)(x + y - 1)\, dx\, dy$$

$$= \frac{k}{8} \int_{-1}^{1}\int_{-1}^{1} \begin{array}{l}(48 \text{ terms, comprising } -1 + 2\,x^2 + y^2 - x^4 - x^4 y^2, \\ \text{and others which give zero integrals})\end{array} dx\, dy$$

$$= -2k/15.$$

Even a single term of **K** takes us a long time. In the bad old days, sadistic teachers used to give their students many, many such exercises.

$$
\mathbf{K} = \frac{k}{45}
\begin{bmatrix}
6 & -6 & 2 & -8 & 3 & -8 & 2 & -6 \\
-6 & 32 & -6 & 20 & -8 & 16 & -8 & 20 \\
2 & -6 & 6 & -6 & 2 & -8 & 3 & -8 \\
-8 & 20 & -6 & 32 & -6 & 20 & -8 & 16 \\
3 & -8 & 2 & -6 & 6 & -6 & 2 & -8 \\
-8 & 16 & -8 & 20 & -6 & 32 & -6 & 20 \\
2 & -8 & 3 & -8 & 2 & -6 & 6 & -6 \\
-6 & 20 & -8 & 16 & -8 & 20 & -6 & 32
\end{bmatrix}
$$

the term
calculated
above

It would almost be easier to write a program. The problem to be solved is shown in Fig. 4.3, and we assemble **K** to give Table 4.1. As before, only more tedious. For the loads on the right-hand side of the equilibrium equations, we take the total weight of 4 for each element, and divide it evenly amongst the eight nodes, as before.

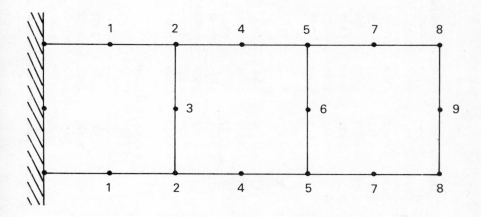

Fig. 4.3 – Plan view of a membrane on an elastic foundation. This exercise is roughly comparable with that of Table 3.1 and Fig. 3.2(b).

Table 4.1.

The full solution for Fig. 4.3. This matrix contains a wider band of nonzero coefficients than in the previous equations.

Original coefficients

w_1	w_2	w_3	w_4	w_5	w_6	w_7	w_8	w_9	RHS	Check sum
2.133	−0.622	0.889	0	0	0	0	0	0	1.000	3.400
−0.622	0.711	−0.533	−0.622	0.222	−0.356	0	0	0	2.000	0.800
0.889	−0.533	1.422	0.889	−0.356	0.356	0	0	0	1.000	3.667
0	−0.622	0.889	2.133	−0.622	0.889	0	0	0	1.000	3.667
0	0.222	−0.356	−0.622	0.711	−0.533	−0.622	0.222	−0.356	2.000	0.666
0	−0.356	0.356	0.889	−0.533	1.422	0.889	−0.356	0.356	1.000	3.667
0	0	0	0	−0.622	0.889	2.133	−0.622	0.889	1.000	3.667
0	0	0	0	0.222	−0.356	−0.622	0.356	−0.267	1.000	0.333
0	0	0	0	−0.356	0.356	0.889	−0.267	0.711	0.500	1.833

REDUCED COEFFICIENTS (above and on the diagonal) and **MULTIPLIERS USED DURING REDUCTION** (below the diagonal)

w_1	w_2	w_3	w_4	w_5	w_6	w_7	w_8	w_9	RHS	Check sum
2.133	−0.622	0.889							1.000	3.400
−0.292	0.529	−0.273	−0.622	0.222	−0.356				2.292	1.793
0.417	−0.516	0.910	0.568	−0.241	0.172				1.766	3.174
	−1.175	0.624	1.048	−0.211	0.363				2.591	3.793
	0.419	−0.265	−0.201	0.512	−0.265†	−0.622	0.222	−0.356	2.028	1.518
	−0.672	0.189	0.347	−0.518†	0.887	0.567	−0.241	0.172	2.358	3.742
				−1.216	0.639	1.014	−0.198	0.346	1.958	3.122
				0.434	−0.272	−0.195	0.155	0.002	1.143	1.301
				−0.696	0.193	0.341	0.011	0.312	0.776	1.089

w_1	w_2	w_3	w_4	w_5	w_6	w_7	w_8	w_9		
2.031	7.396	1.426	2.972	6.890	2.563	2.516	7.342	2.487	ANSWERS	
3.033	8.399	2.423	3.974	7.889	3.562	3.518	8.349	3.490	CHECK	

† With a large 'bandwidth' as here, you can nearly halve the number of arithmetical operations without jeopardizing the checks, by observing that for example −0.518 = −0.265/0.512. Calculate the future multipliers, as you write the coefficients. The computer would not even store the numbers below the diagonal.

4.3 IMPASSE: NODAL LOADS

Curse! All that hard work ... pesky elements. A chapter about 'good' elements? These are the *worst* answers so far! The deflections in Table 4.1 are not only erratic, but much too high.

Or are they? Hold it. At the centre of the element, N_F or any other midside shape function is $\frac{1}{2}$, while N_E or any other corner function is $-\frac{1}{4}$, *negative*. Yes. Thus from Table 4.1, the *centres* of the elements have deflections $-0.954, -2.177$ and -2.075, strongly negative.† Dare we hope? The mean deflections may be about right! Perhaps the distribution of the forces is very wrong?

It is indeed. The elements are distorting horribly. Now do a thought-experiment. Imagine an isolated element, with uniform weight and uniform spring support. In this simple case, every point — including the nodes — surely *must* descend by the same amount? Yes! Now work backwards. We have **K**, the element stiffness, two paragraphs up. *Assume* that $w_A \dots w_H$ are all unity. Multiplying out gives the nodal forces that would have produced this result, $(-\frac{1}{3}, \frac{4}{3}, -\frac{1}{3}, \frac{4}{3}, -\frac{1}{3}, \frac{4}{3}, -\frac{1}{3}, \frac{4}{3})$. We want *negative* loads at the corners! Remarkable.

This thought-experiment is all very well; but suppose the loading had been non-uniform. Whenever we need a nodal load, we must do a thought-experiment, but a slightly different one. Try to think about the shape function constraints.‡ They do three jobs: (i) they connect the nodes to the springs (distributed as here, or lumped, it makes little difference); (ii) they connect neighbouring elements, all along the edges; and (iii) they connect the *external loads* to the springs (in this case, gravity-loads). Or, we now argue, they connect the external loads to the *nodes*.

We want to find the loads X at the nodes which would produce the *same* overall effect as the gravity-loads. There is a general way to approach this type of problem, known as 'virtual work'. See Fig. 4.4. For example, let us make a tiny change at node A; let us increase w_A to $w_A + dw_A$. Then we can find how much work gravity would do.

---

† For example, $-0.954 = -\frac{1}{4}$ (four corner values) $+ \frac{1}{2}$ (four midside values)

$$= -\frac{1}{4}(0 + 0 + 2w_2) + \frac{1}{2}(0 + 2w_1 + w_3)$$

where $w_1 = 2.031, w_2 = 7.396, w_3 = 1.426$ from Table 4.1.

‡ Admittedly this is difficult in the case of the 8-node element. Simple levers, like needles or tubes, cannot do the job. The idea begins to get fuzzy and abstract. But try to hang on to it.

This must equal the work that the *equivalent* load X_A would have done, namely $X_A \, dw_A$.

Fig. 4.4 – Virtual work. Fiona sits twice as far away from the fulcrum as Penelope. When the see-saw tips $d\theta$, F goes down twice as far as P goes up. In total, the work done by gravity cancels. This is what moment-equilibrium is all about.

Thus a piece of the membrane remote from A has weight $L \, dx \, dy$, and dw_A makes it descend a smaller amount, $N_A \, dw_A$. The total work done by gravity is

$$dw_A \int_{-1}^{1} \int_{-1}^{1} N_A L \, dx \, dy = X_A \, dw_A \, .$$

So
$$X_A = \int_{-1}^{1} \int_{-1}^{1} N_A L \, dx \, dy$$

$$= \int N_A \text{ times (little force) over element}$$

$$= L \int_{-1}^{1} \int_{-1}^{1} \tfrac{1}{4} (1 + x)(1 + y)(x + y - 1) \, dx \, dy$$

$$= -\tfrac{1}{3} L \, .$$

The midside forces are even easier. No problem remains. Let us now repeat the exercise of Fig. 4.3, this time with the correct loads: see Table 4.2 ... Ah, that's better. Heaven is in sight. The reduced

Table 4.2

As Table 4.1, but with the correct nodal loads as determined by virtual work. Observe that many are now negative.

w_1	w_2	w_3	w_4	w_5	w_6	w_7	w_8	w_9	RHS	Check sum
2.133	-0.622	0.889	0	0	0	0	0	0	2.667	5.067
-0.622	0.711	-0.533	-0.622	0.222	-0.356	0	0	0	-1.333	-2.533
0.889	-0.533	1.422	0.889	-0.356	0.356	0	0	0	2.667	5.334
0	-0.622	0.889	2.133	-0.622	0.889	0	0	0	2.667	5.334
0	0.222	-0.356	-0.622	0.711	-0.533	-0.622	0.222	-0.356	-1.333	-2.667
0	-0.356	0.356	0.889	-0.533	1.422	0.889	-0.356	0.356	2.667	5.334
0	0	0	0	-0.622	0.889	2.133	-0.622	0.889	2.667	5.334
0	0	0	0	0.222	-0.356	-0.622	0.356	-0.267	-0.667	-1.334
0	0	0	0	-0.356	0.356	0.889	-0.267	0.711	1.333	2.666

REDUCED COEFFICIENTS / MULTIPLIERS USED DURING REDUCTION

w_1	w_2	w_3	w_4	w_5	w_6	w_7	w_8	w_9	RHS	Check sum
2.133	-0.622	0.889							2.667	5.067
-0.292	0.529	-0.273	-0.622	0.222	-0.356				-0.554	-1.053
0.417	-0.516	0.910	0.568	-0.241	0.172				1.269	2.678
	-1.175	0.624	1.048	-0.211	0.363				1.224	2.426
	0.419	-0.265	-0.201	0.512	-0.265	-0.622	0.222	-0.356	-0.519	-1.028
	-0.672	0.189	0.347	-0.518	0.887	0.567	-0.241	0.172	1.361	2.746
				-1.216	0.639	1.014	-0.198	0.346	1.166	2.329
				0.434	-0.272	-0.195	0.155	0.002	0.156	0.313
				-0.696	0.193	0.341	0.011	0.312	0.310	0.623

w_1	w_2	w_3	w_4	w_5	w_6	w_7	w_8	w_9	
1.149	0.833	0.827	1.027	0.969	0.969	1.005	0.994	0.994	ANSWERS
2.150	1.827	1.826	2.030	1.971	1.969	2.005	1.994	1.997	CHECK

coefficients are unaltered, but it is as well to repeat the checks. In these exercises the most extraordinary things seem to happen! But clearly it is important to get the right loads at the nodes. If not, how can we expect elements to be good?

Of course for real elements, in real programs, we should do this too by numerical integration. At each Gauss point we place a 'representative weight' of L det $\mathbf{J}$, which produces a nodal force $N_A L$ det $\mathbf{J}$ for example. We sum this over the four Gauss points. Easy, for the computer.

4.4 THE 8-NODE ELEMENT IN PLANE STRESS

In real jobs, the 8-node element is one of the most attractive. Not only does it give good answers; but the human labour involved is minimal, which is not immediately obvious. Of course, this is crucial if one's business affairs are to prosper!

Suppose that a job has some curved boundaries. With a straight-edged element, we have to approximate the curves rather carefully, which wastes time. With a curved element – as this can be – we write the coordinates as they actually are!

But many boundaries are straight, and the internal edges are normally straight for simplicity. FEMSKI has a special feature, so that we do not have to specify perhaps 60% of the nodes. If a node without coordinates is a midside, then FEMSKI places it exactly halfway between the two corners. Conceivably FEMSKI could have invented the midside node numbers as well! – but tabulating the coordinates is by far the most arduous task.† Although designed for undergraduate use, the package is sophisticated. It has to be, to provide the diagnostics. We are not anxious to add further complications, unless the gain is very substantial.

An example chosen to show the power of this element is shown in Fig. 4.5. Because the vertical tensions cause horizontal contractions, this problem is not as simple as it looks. There is a very local infinite stress at the top right-hand corner! The results give some hint of this. Do this as an exercise, or something like it.

---

† If the coordinates for a *corner* node are unspecified, FEMSKI gives an error message. FEMSKI is in standard FORTRAN IV but with dynamic dimensioning. This is why your teacher need not alter the source code, for everybody who submits a very nonstandard job. It is also why file 5 contains a rudimentary dimension card, for example.

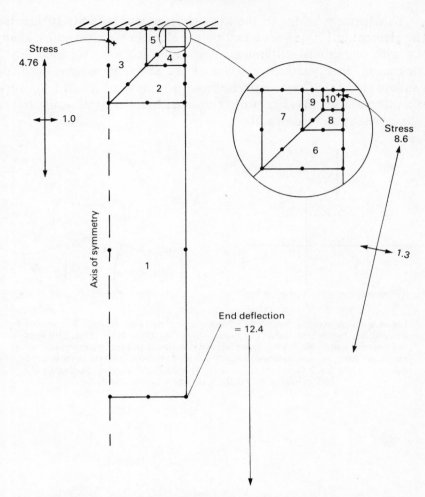

Fig. 4.5 – The hanging signpost again, with ten 8-node elements. There is an infinite stress at the top right-hand corner, but it would require an even finer mesh to show it.

4.5 A DELINQUENT EXAMPLE

Figure. 4.6 shows a job submitted in good faith by a student, many miles away.† The stresses were better than we would expect. But a devil had gotten into the displacements.

† Thanks to D. G. Naylor, Swansea University, UK.

This forms a bridge to the next chapter, on 'Naughty Elements'. The element in FEMSKI is based on the element we worked by hand, but with one crucial difference. Yes, we misled you. We used 2 × 2 integration. In consequence some of the terms are wrong, even for a square element. These errors improve the answers! (apart from very exceptional cases like Fig. 4.6). You must never trust us again. Never believe what you read in a book.

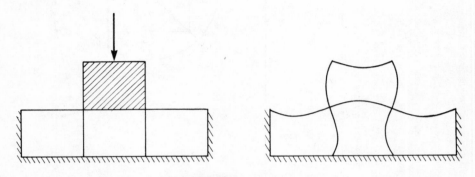

Fig. 4.6 – This exercise represents a forging operation, very crudely. The shaded element has Young's modulus a thousand times that in the other three. This does not prevent a near-mechanism however: what happens is a crazy consequence of our shape function assumptions. When the shaded element deforms as shown, the material at the 2 × 2 Gauss points rotates, with hardly any strain. So despite the high modulus, very little strain energy is generated.

CHAPTER 5

What do naughty elements do?

5.1 BENDING THE RULES

For many years, finite element people all obeyed the rules like angels. It was the easiest way! But gradually the market became more competitive, and we found that misdemeanours sometimes pay dividends ... Some of us began to ask, 'What are these rules anyway?' And 'Who made them?'

It is not *really* a moral question, how to optimize an industrial process – except when human welfare is involved. All that we need consider in this chapter is the quality of the answers, and the risks that we foresee in other jobs, not yet run.

5.2 TRIANGLE WITH MIDSIDE NODES

Although the message is serious enough, do not take our first naughty element too seriously. By putting all the nodes at the midsides in Fig. 5.1, we create a very trivial 'finite element' problem. For the springs are at the midsides too. The weight L of an element may be shared among the midsides instead of the corners, $\frac{1}{3} L$ at each. So the 'perfect' answers of Fig. 5.1 require no calculation at all!

But are they really so perfect? At A, for example, two of the triangles give $w = 1$, the other two $w = 0$. At B or C, two of the triangles give $w = 1$, the third $w = 2$! Many competitive elements allow discontinuities. Does it matter?

The case is quite unlike Fig. 2.8. These individual elements are not 'wobbly'. And if their behaviour is unruly at A, B, and C, it really is *perfect* to the right of BC. Is this not better? Here at least the *computer* is not talking nonsense! Remember this element. But don't take it seriously.

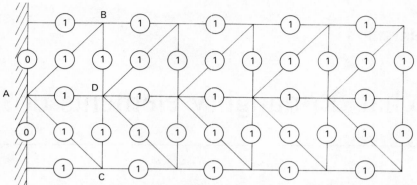

Fig. 5.1 – The membrane of Fig. 2.8(b) reformulated with midside nodes. The ringed numbers are the deflections, with $L = k = 1$.

5.3 2 × 2 INTEGRATION FOR THE 8-NODE ELEMENT

The 8-node square becomes an isoparametric element, if we base the geometry too on the shape functions, $N_A \dots N_H$, as in the Taig quadrilateral. Useful, because then the edges can be curved. See section 4.4. Rumour has it, that the answers are much better with 2 × 2 integration, instead of exact integration as in section 4.2. See section 3.5. Certainly Fig. 4.5, and before that Fig. 2.5, were very good. The stiffnesses are wrong, but the answers are better? Oh.

This again can be a manual exercise, for the square membrane element. Each term in the stiffness matrix in section 4.2 was the integral of the product of two shape functions, both quadratic. Where x^4 or y^4 contribute to K_{EF} for example, we must now replace 2/5 (the exact answer) by 2/9, as given by the Gauss 2-point rule, eq. (3.2). The other terms are exact. Thus with little extra work, we can find the element stiffness matrix with 2 × 2 integration:

$$
\mathbf{K} = \frac{k}{27}
\begin{bmatrix}
2 & -2 & 0 & -4 & 1 & -4 & 0 & -2 \\
-2 & 16 & -2 & 12 & -4 & 8 & -4 & 12 \\
0 & -2 & 2 & -2 & 0 & -4 & 1 & -4 \\
-4 & 12 & -2 & 16 & -2 & 12 & -4 & 8 \\
1 & -4 & 0 & -2 & 2 & -2 & 0 & -4 \\
-4 & 8 & -4 & 12 & -2 & 16 & -2 & 12 \\
0 & -4 & 1 & -4 & 0 & -2 & 2 & -2 \\
-2 & 12 & -4 & 8 & -4 & 12 & -2 & 16
\end{bmatrix}
$$

There are no new problems in creating the equations; the solution progresses much as before (Tables 4.2 and 5.1) – except that the

Table 5.1.

An attempt to solve Fig. 4.3 terminates prematurely, with a null equation.

w_1	w_2	w_3	w_4	w_5	w_6	w_7	w_8	w_9	RHS	Check sum
1.778	−0.444	0.889	0	0	0	0	0	0	2.667	4.890
−0.444	0.296	−0.296	−0.444	0.074	−0.296	0	0	0	−1.333	−2.443
0.889	−0.296	1.185	0.889	−0.296	0.296	0	0	0	2.667	5.334
0	−0.444	0.889	1.778	−0.444	0.889	0	0	0	2.667	5.335
0	0.074	−0.296	−0.444	0.296	−0.296	−0.444	0.074	−0.296	−1.333	
0	−0.296	0.296	0.889	−0.296	1.185	0.889	−0.296	0.296	2.667	
0	0	0	0	−0.444	0.889	1.778	−0.444	0.889	2.667	
0	0	0	0	0.074	−0.296	−0.444	0.148	−0.148	−0.667	
0	0	0	0	−0.296	0.296	0.889	−0.148	0.593	1.333	

REDUCED COEFFICIENTS

w_1	w_2	w_3	w_4	w_5	w_6	RHS	Check sum
1.778	−0.444	0.889	−0.444	0.074	−0.296	2.667	4.890
−0.250	0.185	−0.074	0.712	−0.266	0.178	−0.666	−1.221
0.500	−0.399	0.711	0.000	0.000	0.000	1.068	2.402
	−2.400	1.001				0.000	0.000
	0.400	−0.375					
	−1.600	0.250					

...impossible to proceed

coefficients are slightly smaller. Strange. In particular, the second diagonal divider (pivot) is only 35% of what it was. A big drop ... .

Ouch! It seemed more flexible, but this is *ridiculous*. The *fourth* reduced equation contains only *zeros*. Go home. Call the doctor ... . Maybe it could be worse though: the right-hand side is zero, too. It has the decency to call for answers like 0/0, indefinite but not infinite. So there are an infinity of possible answers! Like an unloaded mechanism — see section 1.7.

In fact, imagine Fig. 2.6, but without the solid support at the first two nodes. The elements could wobble, one at a time, and it is also possible for many assembled elements to wobble together, as if they had *all* gone to a party ... .

Two further comments. Imagine all four corners going down, the same amount, and the middle of the element going up. We could obviously balance the two effects, so that the Gauss points are all stationary. No strain energy, no spring loads, no forces!

Second comment. In the Taig element, we were smart. There were four variables, and four springs to control them, which is just right. (Not that it would matter much, if we had 'too many'. The computer will never throw out a structure because it is *redundant*.) This time we have 8 variables, and only 4 springs, so an element on its own must have *at least* four mechanisms! Quite a water-bed. In our equations there were 9 variables, and 12 springs ... . No, this is a tricky case. We assumed symmetry about the centreline, so the springs must have the same loading in pairs — there are effectively only 6 springs. So this problem has only 3 mechanisms; but enough to cause our trouble three times.

Phew!

Do not dismiss the question of mechanisms. They are one of the hazards of finite elements. The most sedate people sometimes have them. And often they do not seem to matter. In a more complicated in-plane problem, we shall find that the 8-node element has only one mechanism; and if there are two elements then each prevents the other from wobbling!

5.4 A RESCUE MISSION

Back to the membrane. If we make an assumption, then all is not lost. In Table 5.2 let us *make* $w_3 = w_2$, $w_6 = w_5$ and $w_9 = w_8$; they were in fact equal in Table 4.2. There are now six variables, which is 'just right' with effectively six springs.†

Table 5.2.

A repair job on Table 5.1, constrained to remove any variation along y in Fig. 4.3.

w_1	$w_2=w_3$	w_4	$w_5=w_6$	w_7	$w_8=w_9$	RHS	Check sum
1.778	0.444	0	0	0	0	2.667	4.889
0.444	0.889	0.444	−0.222	0	0	1.333	2.888
0	0.444	1.778	0.444	0	0	2.667	5.333
0	−0.222	0.444	0.889	0.444	−0.222	1.333	2.666
0	0	0	0.444	1.778	0.444	2.667	5.333
0	0	0	−0.222	0.444	0.444	0.667	1.333
1.778	0.444			REDUCED		2.667	4.889
0.250	0.778	0.444	−0.222	COEFFICIENTS		0.666	1.666
	0.571	1.524	0.571			2.287	4.382
	−0.285	0.374	0.612	0.444	−0.222	0.667	1.502
	MULTIPLIERS USED		0.725	1.456	0.605	2.183	4.244
	DURING REDUCTION		−0.363	0.415	0.112	0.003	0.117
1.498	0.009	1.494	0.019	1.489	0.026	ANSWERS	
2.496	1.016	2.488	1.033	2.481	1.045	CHECK	

This suggests an interesting possibility. In Fig. 2.8, and again in Fig. 5.1, we had simple structures, and in both cases all the spring loads were exactly correct. Does this always happen?

Table 5.2 succeeds, and the results are shown in Fig. 5.2. With 2 × 2 integration, the springs are indeed equally loaded! The deflections are crazy, however; even if we added a lot more elements in line, they would behave just as mischievously. There is no 'damping' of the oscillation.

Look again at the performance of the correctly integrated element in Table 4.2 however. The jump causes local misbehaviour, but the elements conspire to damp it down, so that very little noise remains in element III. It looks better − but the spring loads are worse!

---

† If you want to do this on the computer, with element type 1, you can stabilize it by including a tensile stress $\sigma_{yy} = 1$.

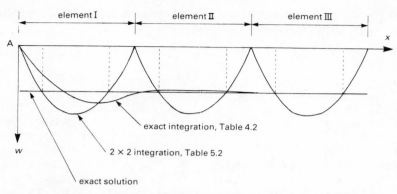

Fig. 5.2 – In both successful solutions, the deflections for Fig. 4.3 do not vary with y, so we can plot them against x. Point A ($x = 0$) is earthed, which is why both curves 'swoop' in this characteristic way.

5.5 BARLOW POINTS FOR SUPERIOR STRESSES?

Many years ago, John Barlow of Rolls-Royce gave birth to an ill-behaved curved beam element. Before dispatching it, he noticed that the bending moments at the two Gauss points in each element were remarkably good, compared with those elsewhere. Since then, Barlow points have been discovered in other elements. They have always been Gauss integration points. The improvement may be tenfold; but there is no guarantee. In the artifical element that we have just seen, the Barlow 'stresses' were perfect!

For the exactly integrated element I in Fig. 5.2, you would hardly call the values at the Gauss points 'optimal'. When the elements interact strongly, and give heavy damping, there is little or no improvement. We have seen two extreme cases, however. In many elements *without* mechanisms or deliberate integration errors, we find a strong oscillation across elements, which damps slowly. Should we expect that they, too, will benefit from John Barlow's unconventional therapy?

Think of the stresses at the Gauss points as 'representative' stresses, somewhat like the representative springs or weights. The engineering instinct is the same.

5.6 LEVELS OF INEBRIATION

Wobbliness, relaxation. We must study it further – it can be a good thing, provided nothing actually wobbles.

Let us think about a variant of Fig. 5.1. Suppose that we had used only one spring, at the centroid of each element, as in Chapter 2. This would have been disastrous. Every centroid would have descended 1. Yes, but this allows the *nodes* far too much freedom. For example, B and C could descend 10 units, and provided D *rises* 7 the springs will give the correct loads.

Isn't it confusing! A small change — we put the nodes at the midsides instead of the corners — and things go from bad to impossible.

Think again. Without the two supported nodes, Fig. 5.1 has 35 variables, 35 degrees of freedom. Suppose that instead of springs, we had a *support* at each centroid: twenty supports for twenty elements. This is not nearly enough for the 35 free variables. Of course it was wobbly. Fifteen times.

'Spurious mechanisms'. (Spurious, lovely word = false, misleading, counterfeit.) These can happen (a) because we do not use enough springs, enough integrating points, or (b) because of *how we join the elements together, where we put the nodes.*

This is subtle. If Fig. 5.1 had corner nodes, there would be only 15 free variables. We seem to have five *redundancies*. Why this enormous change? Coming to think of it, why does the number of variables (including earthed variables) jump from 18 with corner nodes, to 37 with midside nodes? More than double.

5.7 NODAL VALENCY, THE CHEMISTRY OF SUCCESS?

Rough sums. Each triangle has three variables, agreed? Twenty triangles: 3 X 20 gives 60 variables? In both cases? That must be wrong. Try to find the mistake, before reading on.

'Nodal Valency' means the number of elements that connect to a given node. If B and C were corner nodes in Fig. 5.1, their valency would be 3. (As in chemistry: a carbon atom joins to four hydrogens.) The valency of D would be 6. That is, 6 out of the 60 freedoms would be controlled by a single node, by one variable in our equations. We have fewer equations than we might expect. That sounds like a good idea. The 'mean nodal valency' is (the freedoms we expect)/(the freedoms we actually get), 60/37 with midside nodes — and 60/18 with corner nodes, more than 3:1. Quite a saving.

However .... A low 'nodal valency' takes us in the 'mechanism' direction. Provided that we do not actually have mechanisms, this is a much *better* idea. In particular, an individual midside node tends to have a valency of 2, whereas a corner node tends to have a valency of 4, or 6 for triangles. So from our more mature standpoint, it seems that midside nodes are beneficial, in general; not only because we can automate them, and because they enable us to have curved sides, but also because the *performance* should be much better, other things being equal.†

Torgeir Moan first said this clearly. The Barlow points are an easy way to get superlative stresses. But we only expect them to work well, if we can minimize the interaction between elements. When there is a lot of redundancy, this is less likely.

But we soon learn to be cautious, in calculating the level of redundancy, or of wobbliness. Remember simple trusses? We count the tie members, then we count the free variables. If we have more than enough tie members, it *should* be redundant. The snag is, that two members could be doing the same job: or one part of the structure may be redundant, while another contains a mechanism. With finite elements too, it may not be a straightforward count. We conclude this chapter with a somewhat academic discussion of particular cases. It can be quite a tussle.

5.8 SPARRING WITH SATAN

'Good' elements never produce mechanisms unless the real thing is floppy. But 'naughty' elements often carry a jinx. As with trusses, it is obvious − afterwards. See Fig. 5.3 for some examples of a spring-supported membrane. Your input should resemble that in section 2.4, except that the weight per unit area will be unity, and the membrane stress zero.

Now consider the near-mechanism in plane stress, Fig. 4.6. This is more difficult − it anticipates the next two chapters. There are now *three* stresses at each integrating point, σ_{xx}, σ_{yy} and τ_{xy}. It is as if we had three springs, instead of one. With four Gauss points this gives 12 'springs' per element. Figure 5.4 argues that we want 13. Therefore we expect an isolated element to have one spurious mechanism. It has.

On the other hand, Fig. 5.4(d) has 24 springs, and 26 unknowns less the same 3 rigid body motions, which suggests that we have one

---

† This assumes that we have an equation-solver that deals efficiently with midside nodes. Some are not at all good.

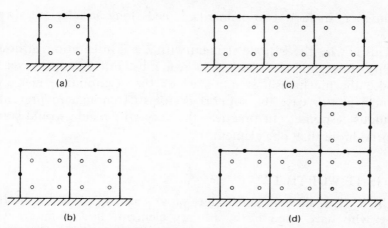

Fig. 5.3 – Mechanisms/Redundancies in spring-supported membranes. The solid dots denote free nodes (one variable, w) and the circles denote supporting springs. In (a) there are too many variables (5) for the springs (4), and we have one mechanism. In (b) there are just enough springs (8). However, one element can misbehave as the mirror image of the other. So the single-element misbehaviour is still possible. In (c) there appears to be a redundant spring, but in fact it is just stable. In (d) there are as many springs as variables, but it fails because the protruding element has a local mechanism as in (a).

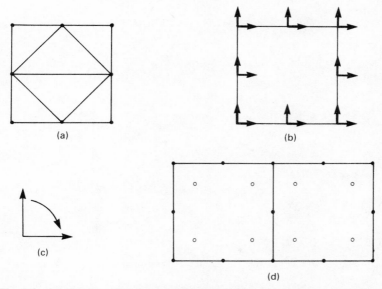

Fig. 5.4 – (a) shows a plane truss which, being a simple structure, represents an 8-node plane stress element which just avoids mechanisms. It has 13 members, which equals 16, the number of variables in (b), less 3, the number of appropriate rigid body motions (c). In (d) two elements are joined, and the spurious mechanism disappears.

redundancy. In fact, two elements joined along an edge are always stable.

This is why the 8-node element with 2 × 2 integration, although naughty, does not give trouble in real jobs. For example, even in Fig. 4.6 the mechanism is stabilized by the neighbouring elements, although in this case the support is not stiff enough to prevent a ridiculous response. (In practice, the very stiff region would surely comprise more than one element?)

5.9 IS IT WORTH IT?

Yes, it is. The answers are better, or you can use coarser meshes. Those who have tried both ordinary elements *and* elements with midside nodes and 2 × 2 integration seem to prefer the latter. However, when you enter the real world of finite elements, it will be wise to generate your *own* experience. See what happens.

CHAPTER 6

How materials deform

6.1 NOT FAR TO GO

So many provocative and intriguing things have happened. And they
have a bearing on what real finite elements do. But it all related to
a trivial case. The computing often gave good answers. But you still
know very little about the *technology* of the real elements. The next
two chapters will rectify this, without too much trouble. For example,
you will discover how easy it is to calculate stiffnesses, in the general
case, thanks to 'strain energy'. Good news.

Stress and Strain? Yes, you did that.† Average stress is force per
unit area – the same units as pressure. Strains are non-dimensional
measures of deformation. We must review some points here, in order
to show how we obtain stresses from finite elements. You need to
appreciate 'principal stresses'. You should already have a vague idea,
from the pictures of stress fields, but this is not good enough.

6.2 TENSILE STRESS AND STRAIN: HOW THINGS STRETCH

If a body is *rigid,* very hard indeed, then no matter how strongly we
load it, a picture drawn on any flat surface would remain the same.

Every real body *deforms* under load, like a lump of putty or of
Plasticine, or of well masticated chewing-gum when you step on it:
although not so much, usually.

The loads are defined as *stresses* (6 of them). We use the symbols

---

† But stress and strain with big deflections (large rotations or strains) is *completely*
different and much more complicated. Remember that.

σ (sigma) for normal (axial) stress and τ (tau) for shear stress†. The deformations are measured in terms of strains (6 of them). We use the symbols ϵ (epsilon) for normal (extensional) strains and γ (gamma) for shear strains.

Let us examine Fig. 6.1. We take as origin the point where a particle on the wire does not shift. We observe that the point x moves to a position $1.2x$ after loading. In finite elements we always speak of DEFLECTIONS – see *Symbols, Definitions* etc. u is the deflection in the x-direction. So here for example we have $u = 0.2x$ for a point originally x to the right of 0. Similarly v would be the movement along $0y$, and w along $0z$. Thus in general $\epsilon_{xx} = \partial u / \partial x$ which we write as u_x. (Formal proof can be found in most books on stress and strain.) Similarly $\epsilon_{yy} = v_y$ and $\epsilon_{zz} = w_z$. Figure 6.2 shows a case which combines ϵ_{xx} and ϵ_{yy}.

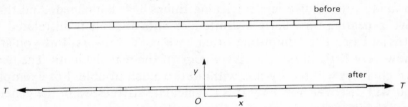

Fig. 6.1 – We mark a fine strip of rubber in cm and then we load it. It stretches from 200 cm to 240 cm. The point 0 does not move. We say the tensile (extensional) strain is 20%, or $\epsilon_{xx} = 0.2$ You remember that extensional strain is change in length divided by original length? Engineering strain, that is.

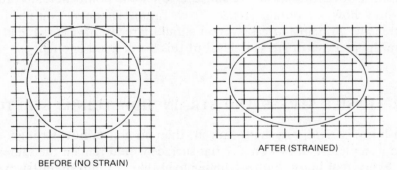

Fig. 6.2 – A sheet of rubber has $\epsilon_{xx} = 0.2$ and $\epsilon_{yy} = -0.2$. Observe that a circle is slightly flattened, to give an ellipse.

† We use the SI unit of stress. 1 N/m² is known as 1 Pascal. To save paper, we should write 1 MPa = 1 Mega Pascal = 10^6 Pascals. We can visualize what so many mega pascals mean physically, if we imagine a wire 1 mm² in section. 1 MPa = 1 N/mm² $\approx$ 1/5 lb/mm². A Pascal is extremely small. Imagine the change in atmospheric pressure when we stand on tiptoe!

6.3 SHEAR STRESSES AND STRAINS: ANOTHER WAY TO CHANGE SHAPE

We *stretch* a wire. We *squeeze* a rubber ball, which is the opposite. When a car *skids*, however, there is a force *along* the surface of a tyre: SHEAR is entirely different. Shear stress is still a force per unit area, but a different force. Figure 6.3 shows us how that third S,

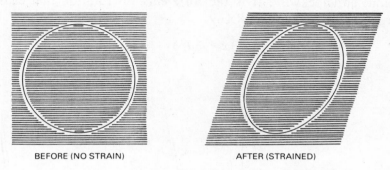

BEFORE (NO STRAIN)   AFTER (STRAINED)

Fig. 6.3 – A pile of papers, neatly stacked, is sheared. Each sheet slips relative to its neighbours. Lines originally vertical are now slanting.

sliding or slipping, transforms a circle into a ellipse at $45°$ to Ox. If we looked at a shear in axes rotated $45°$, we should see only a Squeeze plus a Stretch. Very intriguing. Figure 6.4 introduces the

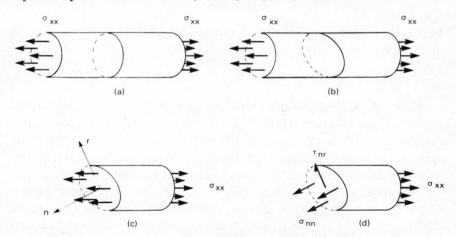

Fig. 6.4 – This bar is under uniform tension. In (a) the stress is 'tensile' across the section at the centre, being normal to it. In (b) it is not, for (c) shows the equilibrium condition. The force on the oblique face resolves in (d) into a normal component (σ_{nn}, tensile) plus a tangential component (τ_{nr}, shear). Not a tensile stress plus another tensile stress.

sort of argument that we shall use in section 6.10. But *never* write $\sigma(\text{tensile}) = \sigma_{xx} \cos \theta + \sigma_{yy} \sin \theta$ or anything like it. *Never* think of *stress* as a vector.

For our purposes, Fig. 6.5 gives a complete explanation of shear strain. As drawn, Ox rotates anticlockwise through angle α and Oy *clockwise* through angle β. If $\beta = -\alpha$, then we would have a pure rotation, anticlockwise. It is the *difference* in rotation that causes shear. Thus the *SHEAR STRAIN*:

$$\gamma_{xy} = \alpha + \beta = \partial v / \partial x + \partial u / \partial y = v_x + u_y$$

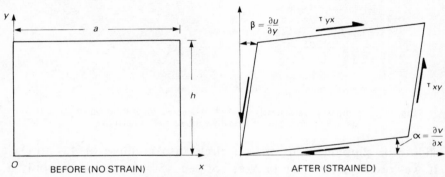

Fig. 6.5 – A rectangle is sheared and rotated at the same time. It is important to distinguish the two effects.

In γ_{xy} the x and y means that it all happens in the xy plane: there is no w. Likewise we have

$$\gamma_{yz} = w_y + v_z \quad \text{and} \quad \gamma_{xz} = w_x + u_z$$

SHEAR STRESSES are relatively easy. With a little explanation, Fig. 6.5 again tells the whole story. There are 'external' forces on the *vertical* as well as the horizontal faces. Indeed, the force per unit area is the same. We say that the shear stress τ_{xy} is equal to the 'complementary' shear stress, τ_{yx}†. There are six stresses, not nine. We remark that $\gamma_{xy} = v_x + u_y$, $\gamma_{yx} = u_y + v_x$, which are again equal. There are six strains too. But in 'plane strain' there are only three strains. It will be useful to relate these to the deflections u and v, in matrix form, as follows.

---

† You should be aware of this. But if not, no hassle. The clockwise moment about the origin is $ah\,\tau_{xy}$ and anticlockwise, $ah\,\tau_{yx}$. A uniform state of stress implies that we have no body forces; so the rectangle is 'in equilibrium'. Proved.

$$
\begin{bmatrix} \epsilon_{xx} \\ \epsilon_{yy} \\ \gamma_{xy} \end{bmatrix} = \begin{bmatrix} u_x \\ v_y \\ u_y + v_x \end{bmatrix} = \begin{bmatrix} \partial/\partial x & 0 \\ 0 & \partial/\partial y \\ \partial/\partial y & \partial/\partial x \end{bmatrix} \begin{bmatrix} u \\ v \end{bmatrix}
$$

6.4 HOW STRESSES CAUSE STRAINS

Wood, bone, etc. are grained — along their length they have strong, stiff fibres. Even skin is very different across the thickness. But the more commonly used man-made materials — concrete, glass, most metals etc. — are not too different in different directions. We call them *ISOTROPIC*.

We also assume in this chapter that the system is *LINEAR ELASTIC*. That is, the deformations are proportional to the loads, and a body returns immediately to its original shape. This is exceptional in the natural world. It is often so in engineering, however, and it is almost always assumed in calculations.

When we stretch a wire, we have 'uniaxial stress', which means there are no other stresses to complicate things. If the material is isotropic, we can easily measure *YOUNG'S MODULUS E*, because $\epsilon_{xx} = \sigma_{xx}/E$. The units of E are the same as those for stress. The numbers are large — even in Ancient British units. Look at Table 6.1: E is astronomic, even to the Druids!

Table 6.1
The elastic moduli of some common materials. (Orders of magnitude only.)

Material	Young's modulus = E			Poisson's ratio = ν
	GPa (Giga Pascals) (= GN/m^2 = kN/mm^2) $10^9 \, N/m^2$	lb/in^2	tons/in^2	dimensionless
Steel	207	30 000 000	13 400	0.3
Aluminium	72.4	10 500 000	4 690	0.33
Concrete	34.5	5 000 000	2 230	0.1–0.2
Wood (along grain)	3.45	500 000	223	anisotropic
Hard rubber	0.0275	4 000	1.79	0.45
Soft rubber	0.00275	400	0.179	0.495

At the same time, the wire shrinks in diameter. See Fig. 6.6. For aluminium, the diameter would decrease a third as much as the length is increased. Thus $\epsilon_{xx} = \sigma_{xx}/E$, $\epsilon_{yy} = \epsilon_{zz} = -\frac{1}{3}\epsilon_{xx}$. A thin concrete wire – imagine that – might shrink by only a tenth. But a strand of rubber shrinks by almost a half. These fractions are called POISSON'S RATIO, ν.

Fig. 6.6 – Close observation reveals that any long, thin object shrinks in cross-section as one stretches it. How much, depends on Poisson's ratio.

By twisting a thin circular tube, we can generate a uniform shear stress. Then by measuring the amount of twist, we can find the *SHEAR MODULUS G*, so that $\gamma_{xy} = \tau_{xy}/G$. However, this tells us nothing new. We could have found G directly from the formula $G = \frac{1}{2} E/(1 + \nu)$. There are only two independent constants, E and ν or E and G.

6.5 MATRICES LOOK TIDIER

If we apply a stress, then another stress, the strains simply add. This property is known as 'superposition'. It only applies to linear elastic systems. Thus

$$\epsilon_{xx} = \frac{1}{E} (\sigma_{xx} - \nu\,\sigma_{yy} - \nu\,\sigma_{zz})$$

$$\epsilon_{yy} = \frac{1}{E} (-\nu\,\sigma_{xx} + \sigma_{yy} - \nu\,\sigma_{zz})$$

$$\epsilon_{zz} = \frac{1}{E} (-\nu\,\sigma_{xx} - \nu\,\sigma_{yy} + \sigma_{zz})$$

$$\gamma_{xy} = \frac{1}{G}\,\tau_{xy}$$

$$\gamma_{yz} = \frac{1}{G}\,\tau_{yz}$$

$$\gamma_{zx} = \frac{1}{G}\,\tau_{zx}$$

(6.1)

We can write this in matrix form. Then we can invert the matrix, giving the *GENERAL FORM*

$$
\begin{bmatrix} \sigma_{xx} \\ \sigma_{yy} \\ \sigma_{zz} \\ \tau_{xy} \\ \tau_{yz} \\ \tau_{zx} \end{bmatrix} = \frac{E}{(1+\nu)(1-2\nu)} \begin{bmatrix} 1-\nu & \nu & \nu & 0 & 0 & 0 \\ \nu & 1-\nu & \nu & 0 & 0 & 0 \\ \nu & \nu & 1-\nu & 0 & 0 & 0 \\ 0 & 0 & 0 & \frac{1}{2}-\nu & 0 & 0 \\ 0 & 0 & 0 & 0 & \frac{1}{2}-\nu & 0 \\ 0 & 0 & 0 & 0 & 0 & \frac{1}{2}-\nu \end{bmatrix} \begin{bmatrix} \epsilon_{xx} \\ \epsilon_{yy} \\ \epsilon_{zz} \\ \gamma_{xy} \\ \gamma_{yz} \\ \gamma_{zx} \end{bmatrix}
$$

$$(6.2)$$

Tedious. But we can easily check the result by finding the product of the matrix implied by equations (6.1) and its supposed inverse in (6.2). This should give the unit matrix:

$$
\begin{bmatrix} 1-\nu & \nu & \nu & 0 & 0 & 0 \\ \nu & 1-\nu & \nu & 0 & 0 & 0 \\ \nu & \nu & 1-\nu & 0 & 0 & 0 \\ 0 & 0 & 0 & \frac{1}{2}-\nu & 0 & 0 \\ 0 & 0 & 0 & 0 & \frac{1}{2}-\nu & 0 \\ 0 & 0 & 0 & 0 & 0 & \frac{1}{2}-\nu \end{bmatrix} \begin{bmatrix} \frac{1}{E} & -\frac{\nu}{E} & -\frac{\nu}{E} & 0 & 0 & 0 \\ -\frac{\nu}{E} & \frac{1}{E} & -\frac{\nu}{E} & 0 & 0 & 0 \\ -\frac{\nu}{E} & -\frac{\nu}{E} & \frac{1}{E} & 0 & 0 & 0 \\ 0 & 0 & 0 & \frac{1}{G} & 0 & 0 \\ 0 & 0 & 0 & 0 & \frac{1}{G} & 0 \\ 0 & 0 & 0 & 0 & 0 & \frac{1}{G} \end{bmatrix}
$$

$$
= \frac{(1+\nu)(1-2\nu)}{E} [\mathbf{I}]
$$

Sometimes we assume *PLANE STRAIN*, which puts $w = 0$ everywhere, so that ϵ_{zz} etc. also disappear:

$$
\begin{bmatrix} \sigma_{xx} \\ \sigma_{yy} \\ \tau_{xy} \end{bmatrix} = \frac{E}{(1+\nu)(1-2\nu)} \begin{bmatrix} 1-\nu & \nu & 0 \\ \nu & 1-\nu & 0 \\ 0 & 0 & \frac{1}{2}-\nu \end{bmatrix} \begin{bmatrix} \epsilon_{xx} \\ \epsilon_{yy} \\ \gamma_{xy} \end{bmatrix} \qquad (6.3)
$$

We ignore σ_{zz} which is nonzero. (Having ignored it, we must not forget it!) Or more often we have a thin sheet loaded only in its own plane, so that $\sigma_{zz} = \tau_{xz} = \tau_{yz} = 0$. Returning to the original six equations, we then have *PLANE STRESS*:

$$\epsilon_{xx} = \frac{1}{E}(\sigma_{xx} - \nu\,\sigma_{yy})$$

$$\epsilon_{yy} = \frac{1}{E}(-\nu\,\sigma_{xx} + \sigma_{yy})$$

$$\gamma_{xy} = \frac{1}{G}\,\tau_{xy}$$

and inverting the 3 × 3 matrix gives:

$$\begin{bmatrix} \sigma_{xx} \\ \sigma_{yy} \\ \tau_{xy} \end{bmatrix} = \frac{E}{1-\nu^2} \begin{bmatrix} 1 & \nu & 0 \\ \nu & 1 & 0 \\ 0 & 0 & \frac{1}{2}(1-\nu) \end{bmatrix} \begin{bmatrix} \epsilon_{xx} \\ \epsilon_{yy} \\ \gamma_{xy} \end{bmatrix} \tag{6.4}$$

Exercise: Find the inverse relation to Eq. (6.3). Explain why the diagonal terms are the reciprocals of those in Eq. (6.4).

6.6 STRAIN ENERGY DENSITY

These relations quickly lead us to a general expression for strain energy, defined again in Fig. 6.7. See also section 3.3. The area of the triangle is:

$$\tfrac{1}{2}\text{ (max. load) times (max. deflection).} \tag{6.5}$$

Let us return to Fig. 6.1, with a wire of length l and cross-sectional area a. If the strain is ϵ_{xx}, then the only nonzero stress is $\sigma_{xx} = E\cdot\epsilon_{xx}$ giving the tension $T = a\sigma_{xx} = aE\cdot\epsilon_{xx}$. The extension δ is $l\cdot\epsilon_{xx}$, and Eq. (6.5) gives:

$$\text{Strain energy} = \tfrac{1}{2}T\delta = \tfrac{1}{2}(a\,\sigma_{xx})(l\,\epsilon_{xx}) =$$
$$\tfrac{1}{2}E\,\epsilon_{xx}^2 \text{ times (volume).} \tag{6.6}$$

The quantity $\tfrac{1}{2}E\epsilon_{xx}^2$ is the *STRAIN ENERGY DENSITY* per unit volume. In all that follows, we shall integrate such expressions over the volume of an element.

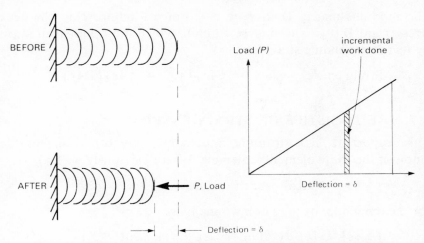

BEFORE

Load *(P)*

incremental
work done

AFTER

P, Load

Deflection = δ

Deflection = δ

Fig. 6.7 – When any spring is compressed, strain energy is stored, equal to the area
under the load-deflection graph.

The energy due to shear is a little more difficult. Let the area
of the paper in Fig. 6.3 be A, and the height h. With a shear strain
of γ_{xy}, the shear stress is $G \cdot \gamma_{xy}$, giving a total skid-force of $AG \cdot \gamma_{xy}$.
Because γ_{xy} is the angle of tilt in radians, the top sheet displaces
$h \cdot \gamma_{xy}$, and the formula gives:

$$\text{Strain energy} = \tfrac{1}{2} AGh \cdot \gamma_{xy}^2 = \tfrac{1}{2} G \gamma_{xy}^2 \text{ times (volume)}.$$
$$(6.7)$$

Before we combine Eq. (6.6) and (6.7) we must check that we
are not doing something foolish. At some stress-level $(\sigma_{xx} \ldots \tau_{xz})$, let
us imagine a small change in all the six strains, $(d\epsilon_{xx} \ldots d\gamma_{xz})$. Then
the work done by the stresses, per unit volume is:

$$\sigma_{xx}d\epsilon_{xx} + \sigma_{yy}d\epsilon_{yy} + \sigma_{zz}d\epsilon_{zz} + \tau_{xy}d\gamma_{xy} + \tau_{yz}d\gamma_{yz} + \tau_{xz}d\gamma_{xz}.$$

That is, *each stress does work only on its own strain.* The six contributions accumulate independently. This follows because x, y, and z are
orthogonal. Each stress implies force in those directions, on various
faces of the unit cube. Each strain implies movements of points on
these faces, in only one of the directions. We have to consider this in
detail to convince ourselves.

We now want a general notation for Eq. (6.2), (6.3), or (6.4):

$$\{\sigma\} = \mathbf{D}\{\epsilon\} \qquad\qquad (6.8)$$

The modulus matrix **D** may be 3 × 3 or 6 × 6, etc. Thus the peak stresses are **D**$\{\epsilon\}$. The matrix formula which multiplies each stress by its corresponding strain is:

$$\text{Strain energy density} = \tfrac{1}{2}\{\epsilon\}^{\text{T}}\{\sigma\} = \tfrac{1}{2}\{\epsilon\}^{\text{T}}\mathbf{D}\{\epsilon\}.$$

6.7 THE LOGISTICS OF STRAIN ENERGY

This important little formula, more than any other, makes life amongst the finite elements tolerable. Where previously we had

$$\int \tfrac{1}{2}\,k\,w^2\,\text{d(area) over the element},$$

for the remainder of the book we shall have

$$\int \tfrac{1}{2}\{\epsilon\}^{\text{T}}\mathbf{D}\{\epsilon\}\,\text{d(volume) over the element}.$$

Thus we have deduced the strain energy for the general 3D case, for plane stress in 2D, and for plane strain. We could go further: we have not mentioned the bending of plates and shells, but these are only engineers' theory of beams in a more general form. Such things are not very interesting, unless one enquires into the molecular/microstructural reasons for the behaviour of materials; this forms a separate course. We have reduced this chapter to the bare essentials.

6.8 MODULUS MATRICES AND THE DEVIL

Our friend makes a close study of materials, too. *He* has fun. Certain values of Poisson's ratio are − amusing. A piece of material with ν exceeding a half will explode at a touch. For this allows negative strain energy, and Nature − always mean with energy − dives straight into the bottomless pit. Ha!

Having been told this, you are unlikely to repeat the mistake. Danger still lurks, however, when the material is anisotropic. In the simplest case there are *five* moduli instead of two. The formulae look very clever, but in fact they are stupid and do nothing to protect you. The only trouble-shooting ploy that we know is to take the modulus matrix **D**, and reduce it, exactly as if we were solving equations. While doing this, we look at the pivots, the diagonal dividers. If they are positive, then all is well.† If not ... don't touch!

------

† A test for positive-definiteness. See any good book on Numerical Analysis.

6.9 PRINCIPAL STRESSES AND LOAD PATHS

We shall spend this laboratory session just getting some insight into principal stresses, without bothering for the moment about the theory. True, we shall prostitute element 9 of the FEMSKI set, an element we take very seriously indeed† – it was intended for real jobs! So if you are entirely happy about principal stresses, then go away and find some more interesting games to play with element 9.

Before starting, however, there are two points to watch –
(i) In Fig. 6.8(a) we count the nodes thus: first, ABCD circulating one face. The rotation ABCD applied to a right-hand screw would cause it to move towards the opposite face. Then E, F, G, and H are adjacent to A, B, C, and D in turn. Disregarding the screw-rule would give a negative volume.
(ii) Figure 6.8(b) suggests an easy way to fix the cube. There are *seven* earthings against the three translations and the three rotations; but the seventh will not inhibit any of the strains that we wish to study.

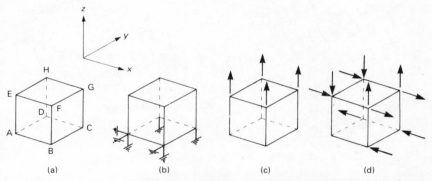

Fig. 6.8 – Experiment, to induce states of uniform stress in an isolated 8-node brick. (a) shows the nodal ordering. (b) shows how to earth it. (c) shows how to generate σ_{zz} = constant. (d) shows how to generate τ_{xz} = constant.

We were born with an 'instinct' for principal stresses. No elaborate explanations were needed when the diagrams in previous chapters showed the equivalent fibres, the load-paths, the stress trajectories, the *direct* interactions between different regions of a body. When nature creates wood or bone, the fibres are presumably in the best directions – 'along the stress' – it is clear to a child what this means.

---

† This element is a very recent analogue of the hybrid quadrilateral that you met in Chapter 1. Previously such analogues had three mechanisms.

When we fashion something out of wood, however, we often get it badly wrong, so that a corner chips off in normal use. Synthetic fibrous laminates, too: we are seldom clever enough, and usually the fibres pull apart long before the tension in the fibres reaches breaking-point, or the layers that we stick together delaminate.

Let us consider two or three *families* of fibres. Suppose that the plastic we use to glue them together is extremely soft. It can pass a small load where the cables of two different families cross, nothing more. Then in Fig. 6.9 for example the 'material' is ultra-selective in the stresses it can accept. But can we design the material to accept *any* given state of stress, by choosing the directions carefully? 'Principal directions'.

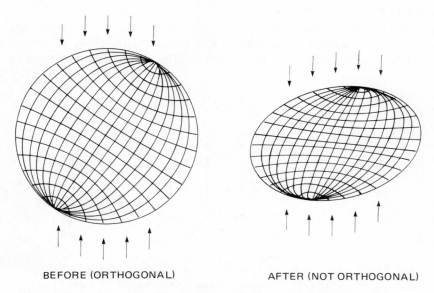

BEFORE (ORTHOGONAL)   AFTER (NOT ORTHOGONAL)

Fig. 6.9 – We consider a hypothetical sphere. If the material consists solely of three families of fibres, mutually orthogonal, and a load is applied in any direction not lying parallel to some of the fibres, then the body collapses in shear.

6.10 BASIC THEORY OF STRESS

'Principal stresses'. As a practical engineer, it is a good idea to know also a little of the mathematics, dealing with stress generally. It was unfortunate that we had to insist in section 6.3 that shear was altogether different from tension. In fact, it is only a different part of the same thing, the 'stress matrix'.

To lead into the stress matrix, we must first consider an old concept in a new light. Often, area can be a vector! Imagine the closed box ORST in Fig. 6.10, containing unit pressure. The pressure loads on the four faces of the tetrahedron are shown as $-\vec{A}_x, -\vec{A}_y, -\vec{A}_z$ and $\vec{A}$. When we pressurize the box, the only extra force that we introduce is the additional weight of air! So the four forces must be in equilibrium. Let us now *define* these vectors, $\vec{A}_x \ldots \vec{A}$, as the 'vector areas' of the faces. It follows that A_x, A_y, A_z (now scalars) are the *x, y,* and *z* components of $\vec{A}$. Evidently the only thing to watch is the *sign* of a vector area. It is the *outward-pointing* normal to a face.

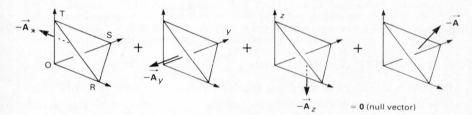

Fig. 6.10 – A tetrahedral container is pressurized. The forces on the faces are in equilibrium. This thought-experiment justifies the concept of area as a vector. See text.

The 3 × 3 'stress matrix' in the equation below is a slightly more mathematical concept. Its only justification is that the equation is true! Accordingly, we multiply the 3 × 3 into the vector area of the sloping face of the box, and then argue that it in fact gives the force on the sloping face, just as the force under unit pressure was $\vec{A}$:

$$\begin{bmatrix} \sigma_{xx} & \tau_{xy} & \tau_{xz} \\ \tau_{xy} & \sigma_{yy} & \tau_{yz} \\ \tau_{xz} & \tau_{yz} & \sigma_{zz} \end{bmatrix} \begin{Bmatrix} A_x \\ A_y \\ A_z \end{Bmatrix} = \begin{Bmatrix} F_x \\ F_y \\ F_z \end{Bmatrix}$$

To prove this formula would require nine pictures like those in Fig. 6.11, some difficult to draw. Instead, we consider only two cases, (a) involving tensile stresses, and (b) involving shear. The other seven contributions to $\vec{F}$ are similar.

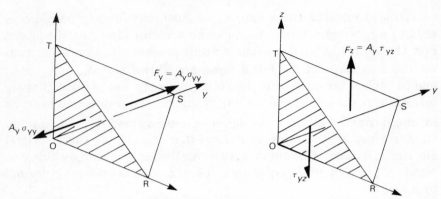

Fig. 6.11 – (a) Only the stress σ_{yy} is present. Accordingly, the face ORT experiences the force $A_y\,\sigma_{yy}$. Faces ORS and OST are unloaded. By equilibrium, the force on the sloping face RST must be as drawn. (b) A similar argument applies, if τ_{yz} is the only nonzero stress. There are also forces $A_z\,\tau_{yz}$ parallel to y, not shown, which would restore moment equilibrium.

6.11 PRINCIPAL DIRECTIONS

In Fig. 6.9, for example, the load-carrying fibres could only carry tension or compression. Fibres cannot carry shear, and the plastic is not firm enough. Let us vary the direction of the face RST $= \vec{A}$ (the normal direction) until the force $\vec{F}(= F_x + F_y + F_z)$ lies precisely along the direction $\vec{A}$. There is thus no shear on the face.

Then let us put one family of fibres along $\vec{A}$, normal to the face. Whatever now happens to the other two families, we can at least be sure that these fibres carry only tension or compression. Writing this idea mathematically,

$$\begin{bmatrix} \sigma_{xx} & \tau_{xy} & \tau_{xz} \\ \tau_{xy} & \sigma_{yy} & \tau_{yz} \\ \tau_{xz} & \tau_{yz} & \sigma_{zz} \end{bmatrix} \begin{Bmatrix} A_x \\ A_y \\ A_z \end{Bmatrix} = \begin{Bmatrix} F_x \\ F_y \\ F_z \end{Bmatrix} = \sigma_i \begin{Bmatrix} A_x \\ A_y \\ A_z \end{Bmatrix}$$

Thus σ_i, a 'principal stress', is still the tension per unit area of cross-section, as in section 6.2. See Fig. 6.12.

In terms of matrix theory, the principal stresses are known as the 'eigenvalues' of the stress matrix, and the principal directions are known at the 'eigenvectors'. This is fortunate, because the subject is very familiar to numerical analysts, and an engineer is unlikely ever to be embarrassed – these devils were tamed long ago. However, you should see a few numerical examples before we quit.

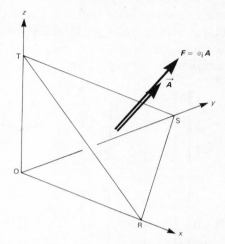

Fig. 6.12 – If $\vec{A}$ happens to lie along a 'principal direction' for the state of stress considered, then $\vec{F}$ is normal to RST. Thus $\vec{F}$ is a scalar times $\vec{A}, \vec{F} = \sigma_i \vec{A}$, where σ_i is the force, normal to RST, per unit area of RST; therefore σ_i is known as a 'principal stress'.

6.12 EIGENVALUES AND EIGENVECTORS

When a matrix is multiplied into its *eigenvector* it gives a scaled replica of the same vector. Here is a simple example:

$$\begin{bmatrix} 1 & 1 & 2 \\ 1 & 0 & -1 \\ 1 & -1 & 1 \end{bmatrix} \begin{bmatrix} 1 \\ 0 \\ 1 \end{bmatrix} = \begin{bmatrix} 3 \\ 0 \\ 3 \end{bmatrix}$$

Scaling factor $= 3$
$=$ 'eigenvalue',
$\sigma_1 = 3$, a principal stress

also:

$$\begin{bmatrix} 1 & 1 & 2 \\ 1 & 0 & -1 \\ 2 & -1 & 1 \end{bmatrix} \begin{bmatrix} -1 \\ 1 \\ 1 \end{bmatrix} = \begin{bmatrix} 2 \\ -2 \\ -2 \end{bmatrix}$$

$\rightarrow \sigma_2 = -2$

and:

$$\begin{bmatrix} 1 & 1 & 2 \\ 1 & 0 & -1 \\ 2 & -1 & 1 \end{bmatrix} \begin{bmatrix} 1 \\ 2 \\ -1 \end{bmatrix} = \begin{bmatrix} 1 \\ 2 \\ -1 \end{bmatrix}$$

$\rightarrow \sigma_3 = 1$

Thus a 3 × 3 matrix has *three* such vectors. These three vectors are always distinct, being orthogonal — their scalar products are zero. But the *eigenvalues* may be equal. They may even be zero: here is another simple example:

$$\begin{bmatrix} 1 & 0 & -1 \\ 0 & 2 & 0 \\ -1 & 0 & 1 \end{bmatrix} \begin{bmatrix} 1 \\ 1 \\ -1 \end{bmatrix} = \begin{bmatrix} 2 \\ 2 \\ -2 \end{bmatrix} \quad \rightarrow \sigma_1 = 2$$

also:

$$\begin{bmatrix} 1 & 0 & -1 \\ 0 & 2 & 0 \\ -1 & 0 & 1 \end{bmatrix} \begin{bmatrix} -1 \\ 2 \\ 1 \end{bmatrix} = \begin{bmatrix} -2 \\ 4 \\ 2 \end{bmatrix} \quad \rightarrow \sigma_2 = 2$$

and;

$$\begin{bmatrix} 1 & 0 & -1 \\ 0 & 2 & 0 \\ -1 & 0 & 1 \end{bmatrix} \begin{bmatrix} 1 \\ 0 \\ 1 \end{bmatrix} = \begin{bmatrix} 0 \\ 0 \\ 0 \end{bmatrix} \quad \rightarrow \sigma_3 = 0$$

Obviously — in uniaxial stress (section 6.4) *two* of the three principal stresses are zero, and in hydrostatic pressure all three are equal.

Exercise: Write the stress matrix for a pressure *P*.

The vector gives the components of $\vec{A}$. We always normalize it, so that the sum of squares is unity, and it becomes the direction cosines of the *normal* to $\vec{A}$, which is the direction of the principal stress! That is, we divide by $(A_x^2 + A_y^2 + A_z^2)^{\frac{1}{2}}$.

Exercise: Explain what it means if we take the negative square root.

6.13 RETROSPECT

We have come a long way from the case of a wire in tension. First there was the question of shear, then there was the question of how materials respond to states of stress. Then, thinking about the ways in which materials usually break, led us to postulate 'reinforcing fibres', aligned in the best directions to convey the loads from one region to another. Returning to the description of stress, and using matrices and vectors, we discovered how to calculate the 'principal directions' in which these mythical fibres should point. So, in a sense, we came full circle, back to a thin wire loaded by forces along its length.

The philosophy ends here, and the engineering begins. What is the best way of displaying the stresses that the computer calculates? We must create pictures which describe how a stress-field resists the loads, and at the same time holds the material together.† Unfortunately, many engineers are content with 'stress-contours'; these merely show which regions are highly stressed, and give us no hint even of how the material is likely to break. There is nothing for the trouble-shooter to get his teeth into. Devil's delight.

† It is not possible simultaneously to make the directions right *and* to space the lines according to the size of the stress. We have compromised, (a) by simply drawing crosses at the Barlow points, so that one has to imagine the rest, (b) by assuming that the stress is constant over subregions of an element, thus avoiding the contradiction, and (c) by treating each element separately, allowing slight errors in orthogonality within elements, and big discontinuities between elements. It is possible to draw continuous stress-trajectories, but then the spacing is not related to the size of the stress. See section 2.3.

CHAPTER 7

Down to business

7.1 ALMOST THERE

Agreed, stress and strain can be a bore. But this is more exciting.
Everything will open up very easily now. The basics of elasticity
from Chapter 6, in matrix form, provide one key. The other key
is the technique for finding the derivatives of the shape functions
in the Taig quadrilateral. We have already seen how to integrate over
the area; to transform the derivatives, we use the same matrix, the
'Jacobian'. Exciting. Or is it?

Yes. Even for the expert this may be an exciting chapter. If he does
not already know it, he will discover a universal 'technique' for solving
the most difficult problems, even those too difficult to visualize. A
programming trick, *SHAPE FUNCTION SUBROUTINES*. These will
give us a truly amazing flexibility.

7.2 THE MATRIX N – SHAPE FUNCTIONS ARE VECTORS

First we must stretch our definitions. For example, $\{N\}$ was a row
matrix, say $[N_A, N_B, \ldots N_H]$ giving the response to unit vertical
motions at each node in turn. But now the motions may not be
vertical. In 3D for example we must let every node move in all three
directions, u, v, w, in turn, giving *three* variables per node. On top
of that, a beam, plate-bending, or shell problem must have slopes as
well, even more variables. It appears that $\{N\}$ will sometimes be a
very long row.

A row? Remember, this is 3D. If we move a node in the x-direction ($u = 1$) the response at a point within the element *might* also be
in the x-direction – but it might not be! And anything can happen
when we *rotate* a node.

In the second place, then, shape functions are now *vectors*. For the 20-node brick, N is a 3 × 60 matrix; 60 because we have u, v, and w at each node, and 3 because each of the actions causes a vector displacement elsewhere, a column of N. The prospect is terrifying only if we imagine doing it by hand. Indeed this is a simple case, because $N_{A,u} \ldots N_{T,w}$ are in a sense scalars: $u_F = 1$ produces a response $N_{F,u}$ *also in the x-direction,* so that

$$
\begin{bmatrix} u \\ v \\ w \end{bmatrix} = \begin{bmatrix} N_A & 0 & 0 & N_B & 0 & 0 & & 0 & 0 \\ 0 & N_A & 0 & 0 & N_B & 0 & \ldots & N_T & 0 \\ 0 & 0 & N_A & 0 & 0 & N_B & & 0 & N_T \end{bmatrix} \begin{bmatrix} u_A \\ v_A \\ w_A \\ u_B \\ v_B \\ \vdots \\ w_T \end{bmatrix} = N \begin{bmatrix} u_A \\ \vdots \\ w_T \end{bmatrix}
$$

Let us now generalize the process for getting gravity loads etc. At some point inside the element, assume that a particle $dx\, dy\, dz$ feels an external force $(X, Y, Z)\, dx\, dy\, dz$. For example, if it was gravity we should have $(0, 0, -\rho g)\, dx\, dy\, dz$ for a material whose mass-density is ρ. But it could be any other body-force, such as a fictional inertia force like centrifugal force etc.

In the general case, the virtual work is the product of the *vector* load into the virtual displacement, another vector. An equivalent nodal load is therefore the *scalar* product of the *vector* shape function into elementary forces, summed over the element: the column of nodal loads is, conveniently,

$$
\int N^T \begin{bmatrix} X \\ Y \\ Z \end{bmatrix} d \text{ (volume)}.
$$

The computer will do such boring tasks, without ever asking for a raise. Just as we had 'representative springs' in the membrane examples, so we have 'representative weights' at the Gauss points which, through the imaginary levers of the shape function constraints, reproduce the effects of the distributed gravity loads etc.† Remember that the big matrix **N** still implies a shape function constraint.

7.3 THE STRAIN-DEFLECTION MATRIX B

The operator matrix, for example the 3 × 2 which relates strains to displacements in plane strain,

$$\begin{bmatrix} \epsilon_{xx} \\ \epsilon_{yy} \\ \gamma_{xy} \end{bmatrix} = \begin{bmatrix} u_x \\ v_y \\ u_y + v_x \end{bmatrix} = \begin{bmatrix} \partial/\partial x & 0 \\ 0 & \partial/\partial y \\ \partial/\partial y & \partial/\partial x \end{bmatrix} \begin{bmatrix} u \\ v \end{bmatrix}$$

is really only a technique of thinking; it is usually nameless. See sections 6.2, 6.3. The important matrix has a symbol, **B**, and we get it by 'operating' on the shape function matrix: the matrix **B** must be present in every program.

$$\begin{bmatrix} \epsilon_{xx} \\ \epsilon_{yy} \\ \gamma_{xy} \end{bmatrix} = \begin{bmatrix} \partial/\partial x & 0 \\ 0 & \partial/\partial y \\ \partial/\partial y & \partial/\partial x \end{bmatrix} \begin{bmatrix} N_A & 0 & N_B & & 0 \\ 0 & N_A & 0 & \cdots & N_D \end{bmatrix} \begin{bmatrix} u_A \\ v_A \\ \vdots \\ v_D \end{bmatrix}$$

$$= \begin{bmatrix} \dfrac{\partial N_A}{\partial x} & 0 & \dfrac{\partial N_B}{\partial x} & 0 & 0 \\ 0 & \dfrac{\partial N_A}{\partial y} & 0 & \dfrac{\partial N_B}{\partial y} & \dfrac{\partial N_D}{\partial y} \\ \dfrac{\partial N_A}{\partial y} & \dfrac{\partial N_A}{\partial x} & \dfrac{\partial N_B}{\partial y} & \dfrac{\partial N_B}{\partial x} & \dfrac{\partial N_D}{\partial x} \end{bmatrix} \begin{bmatrix} u_A \\ v_A \\ u_B \\ \vdots \\ v_D \end{bmatrix}$$

$$= \mathbf{B} \begin{bmatrix} u_A \\ \vdots \\ v_D \end{bmatrix}.$$

† Another case, easier to visualize. Consider an isolated spring-supported membrane element, of uniform weight. Every spring, k det[**J**], supports its own weight, L det[**J**], which descends L/k. Correct. See also section 4.3.

Here we assume a four-node element. If it was an eight-node element for example, subscript D would become H (the eighth letter) and in the program the DO loop parameters 1, 4 would become 1, 8. The strain energy is:

$$U = \tfrac{1}{2} \int \{\epsilon\}^{\mathrm{T}} \mathbf{D} \{\epsilon\} \, \mathrm{d} \, (\text{volume})$$

$$= \tfrac{1}{2} \, [u_{\mathrm{A}} \dots v_{\mathrm{D}}] \; \boxed{\int \mathbf{B}^{\mathrm{T}} \mathbf{D} \mathbf{B} \, \mathrm{d} \, (\text{volume})} \; \begin{bmatrix} u_{\mathrm{A}} \\ \vdots \\ v_{\mathrm{D}} \end{bmatrix}$$

exactly as in section 3.3, except that the integrand was only a column times a row – a matrix, but a very simple one – whereas now it is a triple matrix product.

The element stiffness matrix $\mathbf{K}$ is the item within the rectangle. With numerical integration, $\mathbf{K}$ becomes the sum of several triple matrix products.

7.4 TAIG QUADRILATERAL – HOW TO TAME THOSE DERIVATIVES

The difficulty is not usually in writing the $\mathbf{B}$-matrix, but in finding the derivatives in the appropriate system of axes. In the Taig quadrilateral, for example, it is easy to find:

$$\frac{\partial u}{\partial \xi} = u_{\mathrm{A}} \frac{\partial N_{\mathrm{A}}}{\partial \xi} + \dots + u_{\mathrm{D}} \frac{\partial N_{\mathrm{D}}}{\mu \xi} = -\tfrac{1}{4} u_{\mathrm{A}} (1-\eta) + \tfrac{1}{4} u_{\mathrm{B}} (1-\eta) + \dots,$$

but this is of no direct use in finding $\epsilon_{xx} = \partial u / \partial x$. We must 'transform' the derivatives, and you may have seen the technique in the calculus class. As calculus teachers we do it with symbols. But here we must do it with numbers, which disturbs some people. This is why we choose an example that we can confirm later: see Fig. 7.1.

The first step is to calculate the Jacobian 2×2 matrix, as in section 3.5.

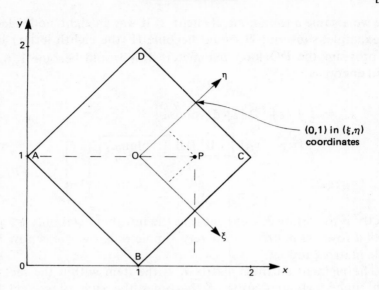

Fig. 7.1 – Example for transforming derivatives in the Taig quadrilateral. The centroid is O, $\xi = \eta = 0$. The selected point is P, $\xi = \eta = \frac{1}{2}$, $x = 1\frac{1}{2}$, $y = 1$.

$$
\begin{bmatrix} \partial x/\partial \xi & \partial y/\partial \xi \\ \partial x/\partial \eta & \partial y/\partial \eta \end{bmatrix} = \begin{bmatrix} \partial N_A/\partial \xi & \partial N_B/\partial \xi \\ \partial N_A/\partial \eta & \partial N_B/\partial \eta \end{bmatrix} \cdots \begin{bmatrix} x_A & y_A \\ \vdots \end{bmatrix}
$$

$$
= \begin{bmatrix} -\frac{1}{4}(1-\eta), \frac{1}{4}(1-\eta), \frac{1}{4}(1+\eta), -\frac{1}{4}(1+\eta) \\ -\frac{1}{4}(1-\xi), -\frac{1}{4}(1+\xi), \frac{1}{4}(1+\xi), \frac{1}{4}(1-\xi) \end{bmatrix} \begin{bmatrix} x_A & y_A \\ \vdots \end{bmatrix}
$$

$$
= \begin{bmatrix} -\frac{1}{8} & \frac{1}{8} & \frac{3}{8} & -\frac{3}{8} \\ -\frac{1}{8} & -\frac{3}{8} & \frac{3}{8} & \frac{1}{8} \end{bmatrix} \begin{bmatrix} 0 & 1 \\ 1 & 0 \\ 2 & 1 \\ 1 & 2 \end{bmatrix} \quad \text{at P}
$$

$$
= \begin{bmatrix} \frac{1}{2} & -\frac{1}{2} \\ \frac{1}{2} & \frac{1}{2} \end{bmatrix}.
$$

Observe that the determinant is $\frac{1}{2}$; the standard 2×2 square in ξ and η transforms into one of half the 'area', in x and y.

Let us now write the chain rule in operator form:

$$\partial/\partial\xi = x_\xi\,\partial/\partial x + y_\xi\,\partial/\partial y$$

$$\partial/\partial\eta = x_\eta\,\partial/\partial x + y_\eta\,\partial/\partial y$$

using as before x_ξ for $\partial x/\partial\xi$ etc. In shorthand matrix form,

$$\begin{bmatrix} \partial/\partial\xi \\ \partial/\partial\eta \end{bmatrix} = \begin{bmatrix} \partial x/\partial\xi & \partial y/\partial\xi \\ \partial x/\partial\eta & \partial y/\partial\eta \end{bmatrix}\begin{bmatrix} \partial/\partial x \\ \partial/\partial y \end{bmatrix}$$

$$= \begin{bmatrix} \frac{1}{2} & -\frac{1}{2} \\ \frac{1}{2} & \frac{1}{2} \end{bmatrix}\begin{bmatrix} \partial/\partial x \\ \partial/\partial y \end{bmatrix}$$

at point P. Remember, this is the matrix we used when we integrated over the area. We took its determinant. This time, we invert it:

$$\begin{bmatrix} \partial/\partial x \\ \partial/\partial y \end{bmatrix} = \begin{bmatrix} 1 & 1 \\ -1 & 1 \end{bmatrix}\begin{bmatrix} \partial/\partial\xi \\ \partial/\partial\eta \end{bmatrix}$$

as in solving a pair of equations. We now have a new composite operator, which we can apply to any of the shape functions, or all of them, taken in turn. This is easy, using matrices:

$$\begin{bmatrix} \partial/\partial x \\ \partial/\partial y \end{bmatrix}[N_A\ N_B\ N_C\ N_D] = \begin{bmatrix} 1 & 1 \\ -1 & 1 \end{bmatrix}\begin{bmatrix} \partial/\partial\xi \\ \partial/\partial\eta \end{bmatrix}[N_A\ N_B\ N_C\ N_D]_P$$

$$= \begin{bmatrix} 1 & 1 \\ -1 & 1 \end{bmatrix}\begin{bmatrix} -\frac{1}{8} & \frac{1}{8} & \frac{3}{8} & -\frac{3}{8} \\ -\frac{1}{8} & -\frac{3}{8} & \frac{3}{8} & \frac{1}{8} \end{bmatrix}$$

(We had already found the ξ–η derivatives at P.)

$$= \begin{bmatrix} \partial N_A/\partial x & \dots & \partial N_D/\partial x \\ \partial N_A/\partial y & \dots & \partial N_D/\partial y \end{bmatrix} = \begin{bmatrix} -\frac{1}{4} & -\frac{1}{4} & \frac{3}{4} & -\frac{1}{4} \\ 0 & -\frac{1}{2} & 0 & \frac{1}{2} \end{bmatrix}.$$

Suppose we had calculated these x and y derivatives of the shape functions at the Gauss integrating points; then we could have entered the values in the strain-deflection matrix **B**, exactly as in the preceding section. It would be easy to create the stiffness matrix.

Exercise: Show that in Fig. 7.1,

$$x = N_A x_A + \ldots + N_D x_D = 1 + \tfrac{1}{2}(\xi + \eta)$$
$$y = N_A y_A + \ldots + N_D y_D = 1 + \tfrac{1}{2}(-\xi + \eta).$$

Using your calculator, find $N_A = \tfrac{1}{4}(1 - \xi)(1 - \eta)$, also x and y, for the points:

$$\left.\begin{array}{l} \xi = 0.45, 0.5, 0.55 \\ \eta = 0.55, 0.5, 0.45 \end{array}\right\} \text{ and } \left.\begin{array}{l} \xi = 0.45, 0.5, 0.55 \\ \eta = 0.45, 0.5, 0.55 \end{array}\right\}$$

(x is constant for the first three, and y for the second.) Hence confirm that for $\xi = \eta = 0.5$, $\partial N_A/\partial x = -\tfrac{1}{4}$ and $\partial N_A/\partial y = 0$, as in column 1 of the final matrix above.

7.5 SHAPE FUNCTION SUBROUTINES – HOW WE WORK TOGETHER

We have been less than honest – some finite elements are difficult to derive. For example, in a shell the derivatives must be in axes in the plane tangent to the shell. One occasionally meets even harder problems than this. So people are beginning to write 'shape function subroutines', which generate the values of the shape functions and their derivatives in the appropriate axes, and also the determinant, and perhaps other things, too. The expert in an element, probably the inventor, writes the subroutine. Then a non-specialist programmer uses the subroutine to create **B** etc. in a larger program. Every element exists with countless variations, in different companies. This way of dividing labour gives the industrial manager all the flexibility that he needs. It seems to suit everybody.

In a book meant for beginners, it is not easy to show how it works out. The remainder of this chapter outlines problems that are increasingly difficult to visualize. These are things you ought to know, as engineers. But we hope at the same time to persuade you that *you* could now go and code the stiffness matrices and nodal force vectors

— given the shape function subroutine. (Which is not to suggest that you should be let loose with even a small program like FEMSKI. Element properties are one thing. Solution routines etc. are quite another.)

7.6 WHAT TO DO WITH ROUND OBJECTS: TORSION

There are plenty of interesting items on the computing agenda this week. The things that we could make with a simple lathe include old-fashioned turned table-legs, wooden eggcups, ordinary drive shafts, discs and flywheels, and frisbees etc. which are much more common than membranes on elastic foundations.

Figure 7.2 introduces the torsion problem. The shape function constraints closely resemble those for the membrane. For the Taig quadrilateral, for example, we could visualize a double infinity of rigid needles, in each radial plane like (c), all around the body. The only strains are γ_{xz} and γ_{yz} because the only deflections are circumferential, in our notation along z.

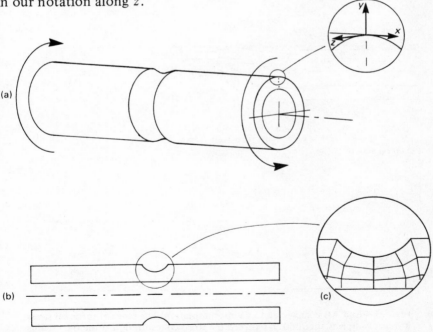

Fig. 7.2 – Torsion of a turned shaft detail. Provided the deformation is small, the deflections are circumferential everywhere, for example 'out of the paper' in (c). Each element represents a ring. Each point represents a circular fibre, which rotates like a rigid hoop. Note, y is the radial direction.

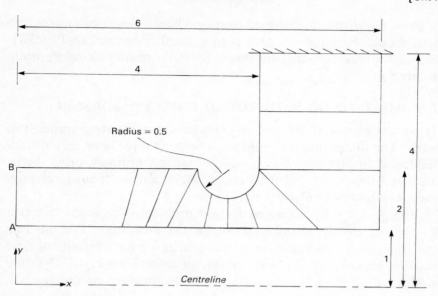

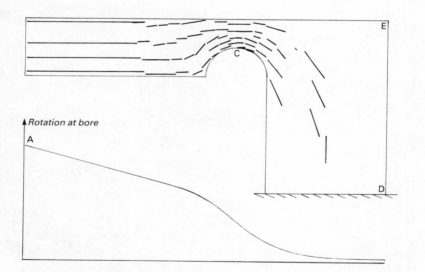

Fig. 7.3 – End AB is subjected to a uniform rotation. The range of torsional stress
in this example is about 60 : 1, between the stress concentration at the fillet radius
C, and the fixing stress at D. We also see an impressive 'stress stagnation' at E,
despite the small radius.

In fact element 3 of FEMSKI includes both the Taig quadrilateral, and the 8-node quadrilateral with 2 × 2 integration. (In this case it never goes wobbly.) Also the corresponding triangles. It is not a very interesting element, however, apart from the concept, and it gives few problems. If you are curious, and you want to try it, remember to fix at least one node! The theory is easy. The strain energy density is

$$\tfrac{1}{2} \begin{bmatrix} \gamma_{xz} & \gamma_{yz} \end{bmatrix} \begin{bmatrix} G & 0 \\ 0 & G \end{bmatrix} \begin{bmatrix} \gamma_{xz} \\ \gamma_{yz} \end{bmatrix}$$

where
$$\begin{bmatrix} \gamma_{xz} \\ \gamma_{yz} \end{bmatrix} = \begin{bmatrix} \partial/\partial x \\ \partial/\partial y - 1/y \end{bmatrix} \begin{bmatrix} N_1 & N_2 & \dots & N_8 \end{bmatrix} \begin{bmatrix} w_1 \\ \vdots \\ w_8 \end{bmatrix}.$$

The volume is slightly trickier:

$$d(\text{volume}) = 2\pi(\text{radius})\, dx\, dy = 2\pi y\, dx\, dy.$$

Outside science fiction, gravity produces no torsional moment. *We* apply the torques. We must interpret a torque by virtual work, rather carefully at first. A deflection dw implies a rotation of dw/y radians (y = radius). Thus if F is the force corresponding to the torque T, we have $F\,dw = T\,dw/y$, or $F = T/y$. (In retrospect, this is obvious. The total load is T/y, evenly distributed around the circumference.)

One example is enough. Figure 7.3 is a sampler, showing various features — the stress patterns that we find in torsion are not too unlike those in tension. But this is really a flow problem. Shear-flow! — with a very subtle difference. Is the *total* shear constant, along a thin shaft of varying radius?

7.7 ROUND OBJECTS UNDER PRESSURE AND TENSION

In industry we would use element 4 of FEMSKI much more frequently. In pressure vessels, flywheels under centrifugal load (constant revs/minute), and rods in tension, there are u and v deflections but w is absent. The strain energy density is

$$\tfrac{1}{2} \begin{bmatrix} \epsilon_{xx} & \epsilon_{yy} & \epsilon_{zz} & \gamma_{xy} \end{bmatrix} \frac{E}{(1+v)(1-2v)} \begin{bmatrix} 1-v & v & v & 0 \\ v & 1-v & v & 0 \\ v & v & 1-v & 0 \\ 0 & 0 & 0 & \tfrac{1}{2}-v \end{bmatrix} \begin{bmatrix} \epsilon_{xx} \\ \epsilon_{yy} \\ \epsilon_{zz} \\ \gamma_{xy} \end{bmatrix}$$

where

$$\begin{bmatrix} \epsilon_{xx} \\ \epsilon_{yy} \\ \epsilon_{zz} \\ \gamma_{xy} \end{bmatrix} = \begin{bmatrix} \partial/\partial x & 0 \\ 0 & \partial/\partial y \\ 0 & y^{-1} \\ \partial/\partial y & \partial/\partial x \end{bmatrix} \begin{bmatrix} N_1 & 0 & \dots & 0 \\ 0 & N_1 & & N_8 \end{bmatrix} \begin{bmatrix} u_1 \\ v_1 \\ \vdots \\ v_8 \end{bmatrix}.$$

This differs only slightly from plane strain, in having the hoop strain ϵ_{zz}.

Exercise: Write **B** and **D**.

There is no dearth of interesting examples using this element. Figure 7.4 shows the typical deflection-patterns that result from

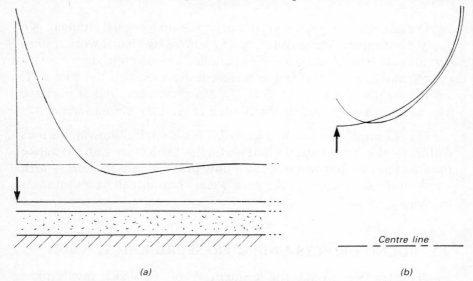

(a)   (b)

Fig. 7.4 – (a) shows the very rapidly damped sine-wave response of a beam on an elastic foundation. A cylinder with a ring-load at the end gives a similar response. (b) shows what may happen in a more complicated shell of revolution (radius 1, thickness = 0.06). The damped sine-wave is again seen, but there is also a distinct bodily shift to the right.

the interplay of hoop stresses and bending effects. In this case, the external load is the true load. In general, for example with pressure, we use the *total* radial or axial load, integrated around the circumference as in torsion.

Figure 7.5 shows an interesting design detail, which does not really need a computer: anybody with engineering commonsense could argue that the spherical pistons should be about as thick as the cylinder. Both have pressure loads only on one side of the circle of contact. And neither has any axial stress at that section.

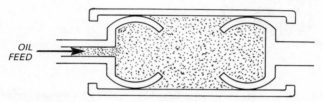

OIL
FEED

Fig. 7.5 – An ingenious spherical piston developed by Rolls-Royce (Derby) for a turboprop torque-meter. The oil pressures range up to about 500 p.s.i. (3450 kPa). The spherical shells are designed to expand exactly as much as the cylinder. Regardless of the pressure, or the orientation of the pistons, there is no increase in the leakage clearance, nor is there a tendency to bind.

Nature makes clever designs too. The largest human joint, shown in Fig. 7.6, has been misunderstood and unjustly criticized! Such designs are of crucial engineering importance when we try to develop artificial replacement joints. If we get it wrong, then a lot of people will suffer.

7.8 ROUND OBJECTS IN BENDING

By far the most interesting element in FEMSKI is number 5, which combines radial, axial and circumferential movements. It represents bending of a shaft etc. and u, v and w have the special interpretation shown in Fig. 7.7. The circles, for example the marks left by the tool, would descend as the shaft sagged; they would also 'tip over', as in the engineers' theory of bending. Although they are stretched and compressed, along their circumferences, they remain circular, to first order, with the same radius.

Observe that v and w are doing almost the same job. In a long thin rod they have to be nearly equal. If not, say $v > w$, there must be a crowding at the top of the circle, that is, a compressive hoop

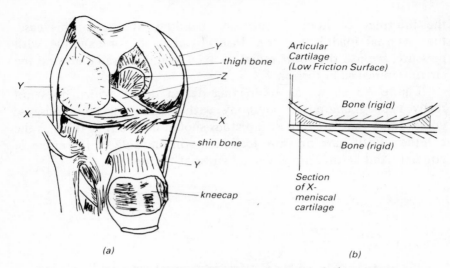

(a)   (b)

Fig. 7.6 — (a) A crude sketch of a flexed knee joint with the kneecap cut away, viewed from the front. The meniscal cartilages (one or both of which are taken out in a cartilage operation) X form the two loops of an '8'. The central part of the '8' is the tibial eminence, a part of the tibia (shin bone) which is 'fixed'. The loops of the '8' are forced outwards, in hoop stress, when we tread heavily. This creates a large contact area between the faces of the femur (thigh bone) with their complex curvature, and the relatively flat faces of the shinbone.

In a healthy person, these surfaces are very slippery. Evidently the '8' can only restrain lateral movement very slightly. The collateral ligaments Y restrain sideways movement and rotation. The cruciate ligaments Z seen near the centre of the joint restrain forwards and backwards movement.

(b) We can model one loop of the '8' tolerably well, as an axisymmetric assemblage of three parts.

strain ϵ_{zz}. This is not at all absurd in a complicated shape, however, nor even in a thin rod. The dominant tensions are along the 'fibres' (or, to please the pundits, the stress trajectories that resist bending — in a thin cone, for example, they are not parallel to the axis). The other stresses being small, we must expect these contractions due to Poisson's ratio, in approximately orthogonal directions like z. Any departure from what we expect would be stiffly resisted, especially in a thin rod.

Thus v and w are almost the same variable. For example, there would be little distinction between the corresponding loads — a strong case of St Venant's principle. Gravity loads go half to one, half to the other.

The shear of a thick cylinder as in Fig. 7.2 would be difficult to visualize. Most of the shear load is carried by τ_{xz} which is greatest at

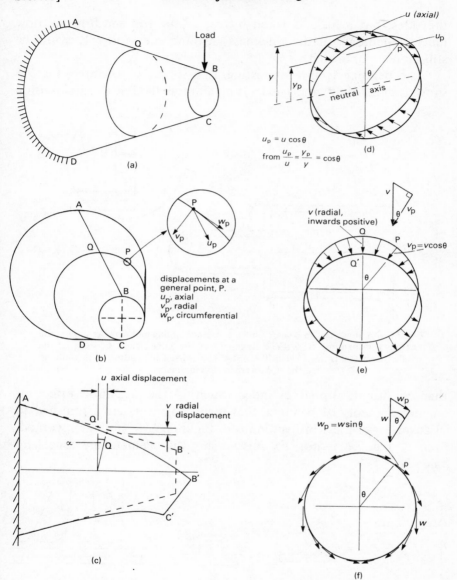

(a)

(d)

$u_p = u \cos\theta$

from $\dfrac{u_p}{u} = \dfrac{y_p}{y} = \cos\theta$

(b)

displacements at a
general point, P.
u_p, axial
v_p, radial
w_p, circumferential

(e)

v (radial,
inwards positive)

$v_p = v\cos\theta$

(c)

u axial displacement

v radial
displacement

(f)

$w_p = w\sin\theta$

Fig. 7.7 – This describes the displacements in bending with shear. The axial deflection u at Q in (c) implies $u \cos\theta$ elsewhere, for example at P, proportional to the height above the neutral axis shown in (d). Therefore u produces the slope associated with bending: the circle does not distort out-of-plane.

According to section (c), v is the downward deflection of Q due to bending and shear. This implies $v \cos\theta$ radially at P, as shown in section (e). But $v \cos\theta$ is proportional to the height above the neutral axis, and is *zero* at the sides. So w supplies *another* 'downward' deflection – actually circumferential as in section (f). This is zero at the top and bottom. To first order, the circle remains a circle, for small u, v and w.

the sides, but a little is taken by τ_{yz} at the top and bottom, how much depending on the thickness. Around the groove it would be still more complicated.

The bending is also interesting. In a shaft, for example, Fig. 7.8 shows a detail that would make it much more flexible in axial tension

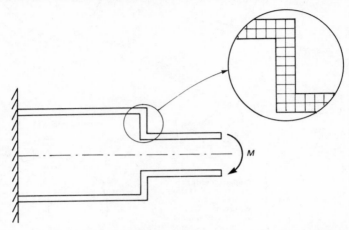

Fig. 7.8 – A 'dog's leg' in a shaft under constant bending moment. As shown, the flexibility in bending is .... The flexibility of the same length of shaft, at the smaller diameter, is .... Evidently the dog's leg is itself in bending, as it would be if the shaft were loaded in tension.

than an uninterrupted cylinder. Such is the hypnotic effect of engineers' theory of bending, that many engineers are astonished to discover a weakness in bending too. Figure 7.9(a) should be worked, even if Fig. 7.8 is not. Exercises using this element are especially valuable.

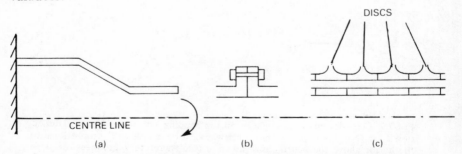

Fig. 7.9 – A case for study, (a) the weakening effect is less marked than in the dog's leg, Fig. 7.8. But note carefully the very high local stresses. This can be disastrous, especially in fatigue. Detail (b) is more common. In (c) many sub-shafts are firmly bolted together. Nevertheless, the contact flexibilities sometimes dominate.

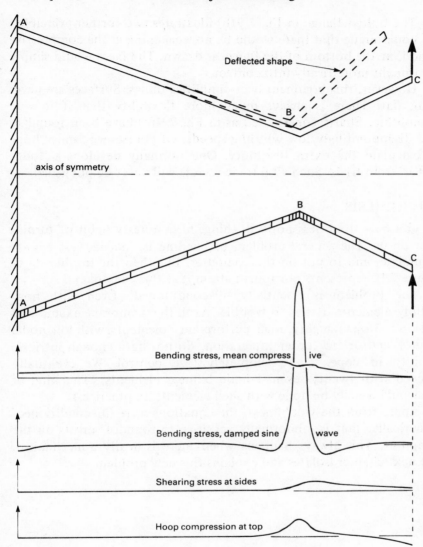

axis of symmetry

Deflected shape

Bending stress, mean compress | ive

Bending stress, damped sine | wave

Shearing stress at sides

Hoop compression at top

Fig. 7.10 – This cantilever, shaped like an old-fashioned milk churn, supports a ring-load at the end, C. Since the generators of the cone AB meet on the line of action of the applied load, the tensions and compressions along the fibres react the shear load as well as the bending moment. This is why there is virtually no shear stress from A to B, apart from transients.

The transient near B accounts for the most damaging stresses, by far. If we loaded the churn in compression, the compressive stresses in the fibres would produce a radially inward ring-load at B. This causes the transient.

The bolted flange in Fig.7.9(b) illustrates two further principles. We could argue that there would be no weakening in the compressive region, at the bottom of the flange as drawn. The faces would simply be brought more firmly into contact.

However, this argument over-simplifies things. Surfaces are never truly flat. There is always much more flexibility than if it were monolithic. Shafts constructed as in Fig. 7.9(c) have been found to give disappointingly low whirling speeds. Of course, we cannot hope to calculate the extra flexibility. One normally develops a 'fudge factor' from trials. Such vital trade secrets will never be published.

7.9 NEMESIS

As you have doubtless noticed, things have a nasty habit of turning sour on us. The general problem of bending is, indeed, very easy to formulate, and to put on the computer. But then the trouble starts. Figure 7.10 represents our fourth attempt at this example.

The problem is essentially 'ill-conditioned'. Even with non-wobbly elements it tries to wobble. As in the membrane example of section 5.3 our naughty, high-performance elements, with low nodal valency and/or low order integration, do not have enough intrinsic stability to cope. They completely lose control. We eventually succeed with twenty-seven 4-node bilinear elements. (This kind of job would usually be done with shell elements, in practice.)

Apart from the wobbliness, the equations were 'ill-conditioned' numerically, that is, abnormally sensitive to roundoff errors on the computer. This always happens when there is nearly a mechanism. The next chapter isolates and explains this new problem.

CHAPTER 8

The hazards of life: when computers go mad

8.1 TAKE A BET?

Of course roundoff is chancy. A microcircuit calculates 2356.3974128309, and passes it on as 2356.397. Umpteen millions of such operations – any one could be disastrous ... don't worry? You try not worrying! Only a fool would not be slightly apprehensive. Roundoff demons everywhere. It *is* a hazard, especially when small computers are entering the scene, with their short wordlengths.

Some structures are intrinsically prone to roundoff. One mesh succeeds, another fails. Our purpose in this chapter is to lead gently into the subject, emphasizing that the main causes are *'physical'*. By the end, you should be able to read the clues. This is much more useful than the rigorous mathematical approach, because we can often do something constructive.

8.2 NEVER TRUST A COMPUTER

Life is full of surprises. Who would have suspected that the simple device shown in Fig. 8.1 would cause so much trouble in the computer-room? Try it and see. In fact, you may have to alter the dimensions, to bring out the worst in *your* computer. Make the small element thicker and shorter, so as to keep the weight constant, and at some stage the deflections and stresses will certainly go berserk. You haven't yet used element 7, the thin beam element, so Table 8.1 might help you.

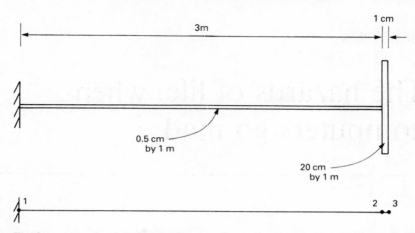

Fig. 8.1 – The patent ultra-springy Irons–Shrive diving board. It is more effective if the thickened section is placed further in-board, so as to trip the faint-hearted *before* they reach the end.

8.3 NEVER TRUST A POCKET CALCULATOR

It is not difficult to understand the principle involved. Let us put two extension springs in series, as in Fig. 8.2. We can write the equations of equilibrium for the two trucks in turn, taking forces towards the left as positive:

$$6001\, u_1 - 6000\, u_2 = 0 \text{ for the left-hand truck,} \qquad \text{(I)}$$

$$-6000\, u_1 + 6000\, u_2 = 1 \text{ for the right-hand truck.} \qquad \text{(II)}$$

To solve these we add 6000/6001 (= 0.99983 rounded off to five places) times the first equation to the second. The computer is stupid, and always works from the top downwards. It so happens that with these simple numbers we can get the exact answers simply by adding the two equations, but in general this is not so. Thus

$$1.02000\, u_2 = 1, \quad u_2 = 0.98039, \quad u_1 = 0.98023 \qquad \text{(II*)}$$

instead of the correct values 1.00017 and 1. Hence the force in the first spring is 0.98023 instead of 1, and in the second spring, 0.96729. We have 3% error in equilibrium. Five-decimal accuracy would seem plenty, with only two equations, but in 'ill-conditioned' jobs like this it is not.

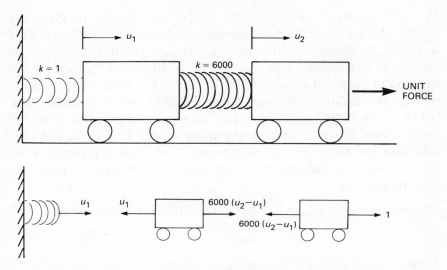

Fig. 8.2 − (a) When a heavy spring is supported by a light spring, the equations of equilibrium tend to be abnormally sensitive to roundoff. (b) The free-body diagrams include the forces due to the springs.

For practical purposes, this is almost as far as we need go. The danger arises when one element is much stiffer than another, when there is something that effectively locks two nodes together with a stiff spring; a stiff lever connecting *three* nodes could be just as bad. And so on.

8.4 NEVER TRUST AN EQUATION

Looking more carefully at equation (II*) we begin to see exactly what went wrong. The number $1.02000 = 6000 - 6000 \times 0.99983 = 6000 - 5998.98000$, working consistently to five decimals. Although 5998.98000 looks very accurate, in fact we can rely on it only to five *significant* figures. What was a small lie, relative to 6000, becomes a 2% lie relative to 0.99983, the accurate value of $6000 - (6000)^2/6001$. The damage is done when we subtract a big number from another big number, and generate a small number, and then divide something by it, for example. Because then it is the *relative* error that matters.

We shall expect you to be familiar with three different ways of quantifying the damage:

(i) The diagonal decay. By the diagonal, we mean the second coefficient in the second equation, the third in the third, and so on.

In finite element equations, the diagonals become progressively smaller during the reduction process (see Table 2.2), until it is time to divide by them: at this stage they remain positive, they may be very small indeed, and they are known as 'pivots'. So in Equation II the overall 'diagonal decay' is 6000 : 1. This is a very rough measure of the proportionate damage done. FEMSKI prints it, sometimes.
(ii) A much better measure is the 'diagonal energy' ratio. The unit force moves almost 1 unit, giving strain energy $U = \frac{1}{2}$. We compare this with the strain energy that would have been generated, if we had imposed the deflections on the diagonal stiffnesses, regarded as individual springs:

$$\text{diagonal energy} = U_D = \tfrac{1}{2}[6001 \times (0.98023)^2 + 6000 \times (0.98039)^2$$

$$= 5766.53$$

Thus $\qquad\qquad U_D/U = 11533.$

Let us try this on the diving-board. Figure 8.3 gives a mental picture of what U_D implies in a real problem. It also suggests why the roundoff problem may disappear if we allow shear distortion in the beam, although this makes virtually no difference to the actual response. For the dimensions shown, the stiffness against lateral deflection, $12\,EI/l^3$ is 1.728×10^9 as great in element 2 as in element 1. Perhaps this will not cause the havoc intended. It depends on the wordlength of your computer. (Actually this example deserves all the contempt that older engineers heap upon the young users of the computers, who fail to consider what their input means *physically*. Engineers' theory of beams is useless, unless the fibres are straight, continuous, and parallel to the axis.)

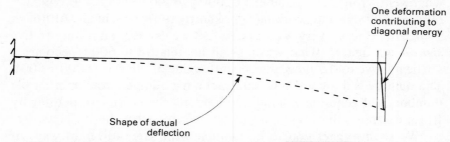

One deformation
contributing to
diagonal energy

Shape of actual
deflection

Fig. 8.3 – An ill-conditioned problem. One nodal variable taken alone gives a pattern of deflections so violent as to give a diagonal energy monstrously greater than the true strain energy. This contribution varies as the inverse cube of the length of the small element, and as the cube of its thickness.

(iii) This time (re-solution) you don't come out with a single number. Never trust an equation. A fertile mathematical concept: when we solve a set of equations with roundoff, we are solving, exactly, the wrong equations! If the equations are 'ill-conditioned' it simply means that the answers are abnormally sensitive to the slightest perturbations, in the coefficients and right hand sides. If 6001 became 6001.01 it would reduce u_1 and u_2 by 1% for example.

Long ago in high school, we used to check our answers by substituting them back into the original equations. In those innocent days, the evidence of a check-substitution was conclusive. Now, with roundoff, the evidence itself is a bit fuzzy. Let's try:

$$0 - 6001 \times 0.98023 + 6000 \times 0.98039 = -0.02023$$

$$1 + 6000 \times 0.98023 - 6000 \times 0.98039 = 0.04000$$

This looks great. Orders of magnitude: if we divide the 'residuals', -0.02023 and 0.04, by 6000 or 6001 — it looks as if the errors in u_1 and u_2 are too small to notice ... Oh? We know that's wrong. Let's solve a *new* set of equations, to see roughly how much we should *really* have to alter u_1 and u_2, to satisfy the original equations:

$$6001\,\delta u_1 - 6000\,\delta u_2 = -0.02023$$

$$-6000\,\delta u_1 + 6000\,\delta u_2 = 0.04000 .$$

Hence $\delta u_1 = 0.01977$, $\delta u_2 = 0.01978$, Because we made no roundoff error in the check-substitution, these corrections are exact.

In general, the roundoffs incurred in the check-substitution are of the same order as (although normally less than) those due to solving the equations. So we cannot always rely on these δu's as corrections. But we can *always* rely on them to tell us, honestly, the 'order of magnitude' of the total roundoff errors.

Do the job properly, of course. Don't stop at the δu. Find what perturbations in stresses, and in support reactions too, these changes imply. Never trust anything.

A more spectacular example should reinforce our message:

$$62x + 67y + 46z = 175$$
$$67x + 73y + 54z = 194$$
$$46x + 54y + 65z = 165 .$$

Check that $x = y = z = 1$. These are the correct answers. Now check our skeleton solution, again using five decimals:

62	67	46	175
1.08065	0.59645	4.29010	4.88625
0.74194	7.19259	0.013830	0.015707

Thus, $x = 1.95477$, $y = 0.023301$, and $z = 1.13572$. Slight disagreement!! Fun, to give this sort of exercise to a class – everybody gets a *different* wrong answer, and the noise is deafening.

With these calculated values, the 'check' RHS, $62x + 67y + 46z$, etc. are 175.00003, 193.99944 and 164.99947, instead of $175, 194$, and 165. Little sense of disaster here – virtually the whole class gets it 'right'.... However, if we re-solve the same equations, with the new 'deficit' RHS, -0.00003, 0.00056, and 0.00053 as before, we get $\delta x = -1.68$, $\delta y = 1.72$, and $\delta z = -0.24$: quite useless for repairing the damage. Just evidence, no more. The δx ... just measure the 'order of magnitude' of the errors. But honestly, this time. The errors are as big as $x, y,$ and z. Disaster.

Evidence is often what is most lacking. This is admittedly an artificial example, created to bemuse our students, but it does emphasize three important facts. (a) The worst 'diagonal decay' is $65/0.013830 = 4700$. You would hardly expect 0.013830 to have nearly 100% error, as it has! Try more decimals here to prove this to yourself. (b) The 'diagonal energy' is $\frac{1}{2}(62x^2 + 73y^2 + 65z^2) = 148$, whereas the strain energy is $\frac{1}{2}(175x + 194y + 165z) = 267$. All okay then? But the diagonal energy only tries to predict errors in the *strain energy*. You should now confirm that the strain energy is indeed almost exact, even if $x, y,$ and z are wildly adrift. (c) On the other hand, re-solution gave the whole unvarnished truth.

FEMSKI does *not* do re-solution, the ultimate in roundoff checking. Not because it isn't important. Rather, because it would lead to various complications, and because to our knowledge no commercial scheme calculates $\delta x, \delta y, \delta z$ ... .

8.5 NEVER GAMBLE

The beams in Fig. 8.4 are of stout English oak, of diameter 20 cm giving a cross-sectional area $A = \pi d^2/4 = 0.0314 \text{ m}^2$, a moment of

Table 8.1

The data file, as created by 'prep' for the example of Fig. 8.4. It has two property lists. Observe the fixing code 111111 – three deflections and three rotations.

How many elements?	How many fix. nodes?	No. loaded nodes?	No. sprung nodes?	No. of R.H. sides?	Maximum R.H. sides?	Iterative solutions?
6	1	1	0	1	1	0

Element	Prop. list	Node numbers					
1	1	1	2	0	0	0	0
2	1	2	3	0	0	0	0
3	1	3	4	0	0	0	0
4	2	2	4	0	0	0	0
5	1	4	5	0	0	0	0
6	1	5	6	0	0	0	0

Property list
Element type 7

Young's modulus	Shear modulus	Area of section	zz-moment (bending)	xx-moment (bending)	Torsion constant	Density z-gravity
3.45e9	0.5e9	0.0314	0.000079	0.000079	0.000157	8390.0

(Integer here and in next line must be in correct column.)
– You mean the very simple beam in three dimensions?

Property list
Element type 7

Young's modulus	Shear modulus	Area of section	zz-moment (bending)	xx-moment (bending)	Torsion constant	Density z-gravity
3.45e9	0.5e9	0.0177	0.000025	0.000025	0.000050	8390.0

(Integer here and in next line must be in correct column.)
– You mean the very simple beam in three dimensions?

Nodes	Fixing-codes	Nodal coordinates		
1	111111	0.0000000	0.0000000	0.0000000
2	0	0.0000000	0.0000000	2.5000000
3	0	0.0000000	0.0000000	4.0000000
4	0	1.0000000	0.0000000	4.0000000
5	0	2.0000000	0.0000000	4.0000000
−6	0	3.0000000	0.0000000	4.0000000

There are no nonzero fixities in this job.

Additional loads						
Node						
5	0.000000	0.000000	736.000000	0.000000	0.000000	0.000000

There are no additional springs to earth.

inertia for bending $I = \pi d^4/64 = 0.000079$ m^4, and a K for torsion = $\pi d^4/32 = 0.000157$ m^4. The strut, element III, has a diameter of only 15 cm giving $A = 0.0177$, $I = 0.000025$, and $K = 0.000050$. The convicted gambler, who now appears to be losing, weighs 75 kg, and hence produces a tension of 736 N in the rope. The data file created by 'prep' is reproduced in Table 8.1, and the answers are also shown in Fig. 8.4.

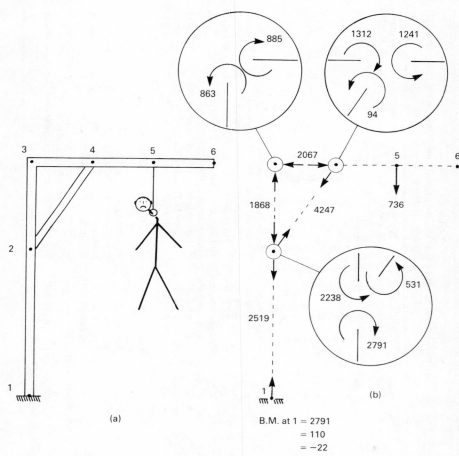

B.M. at 1 = 2791
         = 110
         = −22

Fig. 8.4 – This spectator sport was greatly enjoyed by some in former times. In (a), one of the unhappy contestants is seen hanging from node 5. In (b) the tensions and compressions in the six elements are shown, and the interactions of the bending moments at the joints. Moment equilibrium appears to be violated, by about 20 N.m. This is because the formulation assumes linear variation of bending moment along each element, which tends to be a least-square fit. See the text, however, for devastating comments on this analysis.

8.6 NEVER BOTCH A JOB

A few of these results can be checked by hand. Unfortunately most are wrong. This is because the model was wrong, in assuming rigid connections everywhere. The gallows were not machined from a single piece of metal like some sophisticated modern aircraft structure! The craftsmen would have used loose wooden dowels at the joints – even nails would have been 'exotic'.

We have in fact a simple structure. In our enthusiasm, we have introduced three bogus redundancies. The three joints, at nodes 1, 2, and 4, would definitely not be rigid. Our guess is that they would all be nearly 'pinned', as in the model shown in Fig. 8.5:

(1) Element III (2–4) acts simply as a strut, between nodes 2 and 4. Therefore we make the bending moment zero by putting $I = 0$.

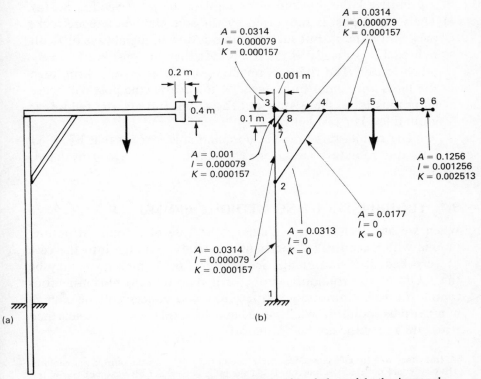

Fig. 8.5 – The extra element, between nodes 9 and 6, models the increased diameter. Every strut element is pinned at both ends, so as to transmit no bending moment. We model a strut with zero moment of inertia I. The triangulation 3–7–8 is an artifice to disconnect the rotation at 8 from that at 3.

(2) Similarly the junction at node 3 is floppy, and transmits no bending moment. But this is more difficult. The FEMSKI input does not allow us to disconnect the rotation θ from one of the elements. So we resort to an artifice, as one often has to. The link 3–8 has zero I: it too transmits no bending moments. Effectively it puts a hinge at the top of the vertical post. It is under tension, and it must have its true area. Link 7–8 carries the vertical reaction, and has the full area of the post.

To avoid duplicating the weight, element 7–3 has a small but finite area. Element 7–3 takes the bending moment, and therefore the horizontal shear force, so it has the full I. (Alone, element 7–3 would not suffice. Without link 7–8, link 3–8 would lack support. We should then have a mechanism.)

(3) The artisan who built the gallows turned the horizontal beam, in a primitive lathe, but did not complete the job.† See Fig. 8.5 (a).

(4) The vertical post is supported by the soil. Maybe there are rocks packed around it. But such features as dowels are always difficult to model realistically. If the soil is soft, then the post is effectively much longer. The fixing is underground. If it is very firm, then the large horizontal forces needed to encaster the post will cause enormous local shear stresses. These will initiate the failure, at ground level or just below.

The results this time are unremarkable, except that FEMSKI puts out roundoff messages. You can't win ... . Do it by hand. See Fig. 8.6.

8.7 TROUBLE-SHOOTING WITHOUT FEMSKI

When we are dealing with numbers, the case of a flying structure, or one with uncontrolled mechanisms, tends to merge into the case of very bad ill-conditioning. See section 1.6. A small pivot may be the focus of the roundoff errors, or it *may* be zero plus numerous roundoff errors. Conversely, a true zero *could* conceivably be a small number plus roundoff. What we see may be similar, with an unearthed structure and with very bad roundoff.

---

† A true story of primitive computers. K.D. submitted a job and the output was rubbish. His boss could not find the bug, nor could the programmer, nor his boss, nor finally the local representatives of the computer manufacturer.

Their Head Office eventually explained it. K.D. had entered the date on his title card, 16:2:61. The computer had taken the colon as an instruction: 'All subsequent calculations to be in pounds, shillings and pence'. These calculations included several square roots.

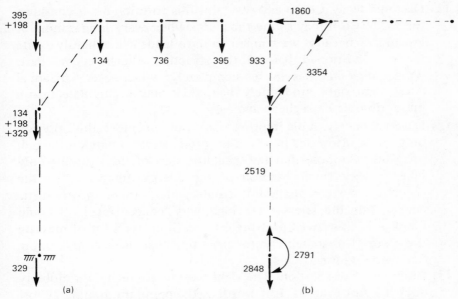

Fig. 8.6 – A hand analysis, following Fig. 8.4. It is now a simple structure, with discrete weights, half of each section at each end of the section as in (a). Compare the results in (b) with those in Fig. 8.4(b) and observe that the stresses are lower, now that the redundancies have gone. Every engineer should know this. When you add a redundancy to a structure, you necessarily make it stiffer. You don't always make it stronger.

If you are running a job on a commercial scheme, with inadequate checks for roundoff, and you feel suspicious about some of the answers, then you have a real problem. Did you go wrong, or did the computer? Probably the best strategy is to estimate the biggest contributions to the diagonal energy, by hand.

Certainly, one recalls with no pleasure at all, his early experiences with roundoff. Typically, the answers looked slightly doubtful. So one did a rough check on the equilibrium. For example, in a turbine disc one checked $\int$(hoop stress) d (area) against (total centrifugal load)$/2\pi$, and found that they differed by 30%. Yet the stresses seemed smooth enough. Roundoff *is* a hazard. Jokes are in bad taste. One's face still turns slightly pink.

8.8 'ROUNDING-OFF'

There are several other ways of measuring roundoff, but they are not worth the trouble of learning. So to recap, we have just three. It's rather untidy, because none of the three is really what we want:

(1) Diagonal decay. Cheap and easy. But this criterion never considers the loads. So it may sound the alarm, when everything is more or less okay – because we happen to have loads which hardly excite the ill-conditioned part of the structure, the 'naughty' part. Worse, it sometimes misses completely: when several diagonal decays are only moderately bad, they may accumulate into a major disaster. It is cheap and easy ... .

(2) Diagonal energy. A big improvement. But remember, this criterion only asks, How badly is the *total* strain energy affected, probably? Suppose that the 'naughty' part of the structure only takes a very small fraction of the energy. Imagine this: the 'naughty' part is small, but troublesome – there is very little energy, but the stresses are high, very 'concentrated'. It could happen. If this trivial part breaks, it will create a lot of mess on the floor, for somebody to clear up. Just as bad as a more 'energetic' failure.

(3) Residuals → re-solution. The right way to go, really. But nobody does it – not even us. Is it worth a 10% premium, to have almost complete security against roundoff? – that is, not to prevent roundoff, but to sound the alarm, if and only if disaster strikes? Evidently most people don't think so, or don't think, or don't know, or don't care, or don't believe what they are told, or naively think they have some magic immunity. But this criterion does involve a lot of extra programming, as well as the 10% premium in running costs. Pity, we all have to learn the hard way.

CHAPTER 9

What drives the user mad?– Spurious side-effects

9.1 RIGOR MORTIS

We have seen (Chapter 2) how a structure goes floppy when there is a spurious mechanism. The elements perform a crazy dance, nothing like the true response. There is little or no strain energy to restrain their exuberance, so the amplitude is random.

We sometimes meet the opposite! The structure almost locks, because certain side-effects generate a great excess of *inappropriate* strain energy. Even the expert may be caught unawares in both cases. But the expert has this advantage: he instantly recognizes the symptoms, and he knows the standard cures. You must learn, too.

9.2 SUCCESS AND FAILURE

In Chapter 1 we saw a good, well-behaved element. The beam in bending, Fig. 1.1, was modelled adequately, using only four elements. You see, each element could model *exactly,* a state of constant bending moment. With a load at the end of the beam, the true solution gives a bending moment that varies linearly, becoming zero at the end. So the hybrid element, type 8, reproduced this picture, to the best of its limited ability (Fig. 1.2) giving a series of equal steps in B.M. each with the same plus and minus errors relative to the true stress field, and correct at the centre of each element.

A good finite element always tries to do this, to be even-handed with the errors. Indeed, if you are lucky enough to know the answers before you start, then there is a rough-and-ready way to choose a suitable mesh. Draw the curve of the stresses. Now imagine that you want to reproduce this curve, by drawing a series of steps, for a constant stress element, or a series of straight lines, for an element giving linear stress variation. In this way, you can usually space the nodes, so as to achieve something like the accuracy that you want.

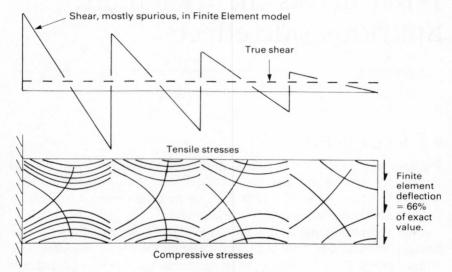

Fig. 9.1 – The same problem as in Figs 1.1 and 1.2. The four bilinear elements give poor deflections, and generally meaningless stress fields. Figure 9.2 provides a qualitative explanation of the poor performance.

Fig. 9.2 – Responses of element type 2 (bilinear) in (a) and of type 8 (hybrid) in (b), to a field of constant bending moment.

In practice, people go through this mental process, but *guessing* what the curve of stress will be. Because it is based on a guess, one draws a finer mesh than one expects to need. It doesn't usually cost much.

Don't count on this happy state of affairs. Occasionally, things go wickedly wrong. Figure 9.1 are the results for a mesh identical to

Fig. 1.1. The answers have changed from Fig. 1.2. Element 2 is the bilinear quadrilateral, described in sections 7.3 and 7.4. Why does it behave so badly?

Simply because the hybrid element in Fig. 9.2(b) is designed to reproduce a state of constant bending moment, exactly. Whereas in Fig. 9.2(a), we only allow the vertical deflection v to vary linearly with x. So the lower edge, for example, has no choice but to remain straight. In consequence, there is a lot of *inappropriate* strain energy, due to the unwanted shear, especially near the ends. This is what makes type 2 much too stiff, in the example of Fig. 9.1.

You might think that nobody uses the bilinear element, but you would be wrong! It is in most commercial schemes. The hybrid is not. The reason, presumably, is that the theory is more difficult; people are frightened more by the theory, than by hybrid elements themselves. But it is enough to know that this hybrid is based on 8 responses, one for each variable $u_1 \ldots v_4$:

3 rigid body motions (up-down, side to side, and rotation).
3 states of constant stress (σ_x, σ_y, τ_{xy}). (These enable it to pass the patch test.)
1 bending response as shown in Fig. 9.2(b).
1 bending response, as part of a vertical beam.
—
8
—

9.3 TOTAL FAILURE

Okay, stupid old world, not to use element 8 rather than 2. But if you want triangles, with corner nodes, then there is no magic hybrid version — and there never will be. Unfortunate: because the situation is really much worse. Figure 9.3 involves 220 lines of data, yet the results are appalling. One reason is that, for illustration, we have chosen a high Poisson's ratio, 0.4, which implies that the material has a high bulk modulus, is relatively incompressible. Even so, you must agree that the results are abominable, considering the human effort involved. This would be a most unkind assignment.†

† Actually we didn't type those 220 lines. We wrote two crude 10-line FORTRAN programs, to generate the node numbers and coordinates. Of course, commercial mesh-generation programs contain more than 10 statements! — but the idea is the same.

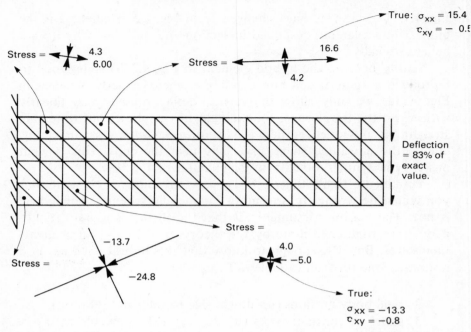

Fig. 9.3 – Hell, not again ... 128 elements, 85 nodes and the answer is 17% low.
The stresses are absurd, too. Figure 9.4 with the accompanying text explains the disaster.

What causes this spectacular failure? Let us consider the single
layer of elements in Fig. 9.4, under constant bending moment.
Elements like A are under horizontal tension, so they want to shrink
vertically. Those like B are compressed, and want to *swell* vertically.
There is a contest, because they are linked by the vertical edges
(except at the ends). Each frustrates the other. When the
sympathetic contractions due to Poisson's ratio are totally prevented,
a material becomes much stiffer in tension — especially if it is nearly
incompressible. So we can explain most of the spurious stiffening
that we observed. This smaller example makes a good assignment,
even with a low Poisson's ratio. The answers are so unexpectedly bad,
that if we hadn't warned you, you might think there was a mistake.
   Figure 9.5 shows an even worse example, also using simple
triangles with a high bulk modulus. This tends to astonish even the
connoisseurs of finite element misbehaviour. For in Fig. 9.5, there
is no bending. You probably remember the case: a spherical soap

bubble for example has equal tension in all directions, and the same principle is used in lightweight pressure vessels, often made of Fibreglass. If you want to know why the finite elements behave so badly here, consult *Techniques of Finite Elements,* p. 145.

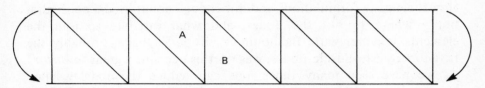

Fig. 9.4 – A fragment of Fig. 9.3, illustrating the malfunctioning of the linear triangle, in a case of plane strain with moderately high Poisson's ratio, i.e. a fairly large bulk modulus.

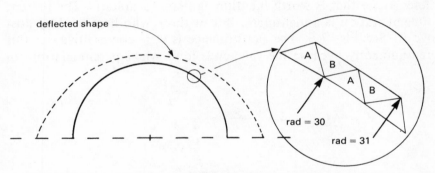

Fig. 9.5 – A thin sphere, with internal pressure, is modelled with 97 linear triangles. Those like A suffer a spurious hydrostatic compression, those like B a tension, amounting to about ± 4% of the true hoop tension with $v = 0.4$.

These fairly small side-effects generate bending moments. Presumably this is why the radial deflections are so spectacularly nonuniform.

Yet the total strain energy is 93% of the correct value.

9.4 LEARN FROM OUR FAILURES

We can draw a different moral for each day of the week. Caution. Preparedness. Astuteness. Murphy's Law. Look for another job. Line up all the computers, and shoot them. Shoot all the researchers, and start again.

Instead, let's think about two less ambitious propositions:

(1) Remember the rectangle? (Fig. 9.1). Unforgivable, but easy to understand. The case of the triangle (Figs 9.3 and 9.5) was different. *Pairs* of elements were involved, misbehaving together. There is no

remedy. Here it is not the 'element formulation' that is to blame, but the *original* choice of nodes. We were dumb to use triangles, with a mean nodal valency of $4\frac{1}{2}$ in Fig. 9.3. (In Fig. 9.1 the nodal valency was only 1.6. Refresh your memory, turn back to section 5.7.)

Even in top finite element society, there is far too much talk about 'element formulation', and not enough about 'nodal configuration' — where to put the nodes, and what variables to join the elements together with. Baudouin Fraeijs de Veubeke, probably the first to suggest midside nodes, was on this account a great *inventor*! As we have seen many times, one can reduce the nodal valency too far. But midside nodes lower the nodal valency from 4 to 2.7 for quadrilaterals, and from 6 to 2.25 for triangles, in a very fine mesh. This is just right, in most cases.

We shall not have time in this course to study shell elements, but there is one that is worth mentioning here, 'Semiloof'. The present implementation is a nightmare. But of those who have used it, most love it. See Fig. 9.6. The performance is very competitive — is this predominantly because the valency is low? And a perennial problem

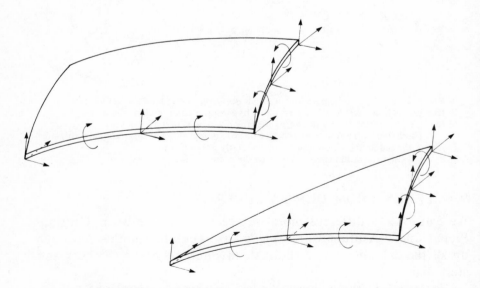

Fig. 9.6 – The Semiloof shell element has 32 variables as a quadrilateral, 24 as a triangle. There are *u, v* and *w* at each corner and midside. In addition, there are the normal rotations, at the sampling points for the 2-point Gauss rule along each side. In a big job, the nodal valency is about 2.5, whereas with all the variables at the corners it would be 4.

with shell elements, how to define the rotations and how to transmit the bending moments, has a simple solution. When somebody manages to reformulate such elements, Semiloof or something like it will probably become the standard shell element.

(2) But the main topic of this chapter, spurious side effects: what is the direct value of this knowledge to the apprentice trouble-shooter? At first sight, not much. If people are dissatisfied with the answers, they will soon switch to a new mesh and/or element, without being prompted. — With one exception. Your boss is a headstrong twit. If he decides to give finite elements a try, and he hits on one of these pathological cases, then he will swear blue murder, and he will vow never to touch finite elements again. One of life's challenges .... Good luck.

CHAPTER 10

Sanity, and shear in beams

10.1 ENERGY AND THE WILL TO CONTINUE

This last chapter has a central message: 'Calculate the energies. Draw conclusions from them. Trust them'. If Chapter 9 shook your confidence, that was our intention. But most of the time things work out surprisingly well. Our strict adherence to energy principles and/or the patch test seems to compensate for our crude shape-function assumptions.† We are backing a winner. So are a lot of other people, with a lot of money.

There are several more interesting elements in FEMSKI, and it is difficult to choose which to end with. We choose to discuss torsion, because with finite elements it has a mind-expanding quality. Physical concepts which are initially quite difficult to grasp, become easy to interpret into finite elements. Without going into details, you will see how we can bypass a lot of – – – – – – – what the mathematicians call 'powerful'. Ugh!

10.2 DON'T BE TWISTED

We saw in section 7.6 what happens when we twist anything that can be turned on a simple lathe – for example, a thick circular tube. Each section of the tube rotates relative to its neighbours, to give shear stress proportional to the radius. The thick tube is as far as an undergraduate course usually takes us, in the English-speaking world. In industry, however, one often meets a beam of complicated section being twisted – aeroplane wings, the coaches of trains, a tall building, etc. See Fig. 10.1 for a description of the torsion bubble.

---

† Even the hybrids are based on energy principles. See Appendix II.

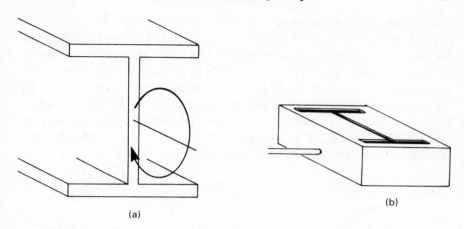

Fig. 10.1 – An I-beam resists a twisting moment in (a), and (b) shows a device for predicting the stress-pattern. A soap-bubble over the I-shaped frame is pressurized from within the box.

The St Venant torsion bubble bulges upwards, and its volume gives the torsional stiffness of the section; its slope gives the shear stress at any point. Experimentally this is a nightmare. The mathematics is obscure. Yet it is on this basis that most people still do their computing. Which we think is a pity. Everybody forgot long ago what it really means physically.

One day, Professor Braine-Rott (whom you must respect, since he is nearly in his dotage) will come and tell you to beware of the dreadful consequences of trusting the computer, of how physical understanding has taken a nosedive in the last twenty years.

This may be true, but it need not be. Torsion, for example; our thesis is that the finite element approach can often clarify a hazy impression left by too much mathematical wizardry. The torsion bubble shows you the consequences, not the cause. You become blasé. You forget where it all started. With beams, you hear so frequently that 'plane sections remain plane', that you begin almost to believe it.

10.3 HOW TO MAKE THINGS WARP

When you twist a rectangular india-rubber ('eraser' if you live in North America) you can *see* the sections warping. Sections do *not* remain plane – see Fig. 10.2(a). Actually, if you want a more formal proof that simple objects deform three-dimensionally when they are twisted, then Fig. 10.2(c) should satisfy you. Let us try to argue that (c) deforms in the same way as (b), but with the sector removed. It would follow that γ_{xy} along the radial face of the groove AC would equal the helix angle to which AC has been twisted, that is, the skewness between AC and the centreline, α in Fig. 10.2(b). Therefore, τ_{xy} should be nonzero. But because there is no surface load, τ_{xy} must be zero. Therefore we are talking nonsense: the section cannot help but warp. Q.E.D.

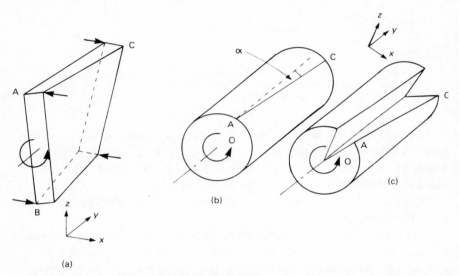

Fig. 10.2 – The rectangular sections of the flat strip (a) remain almost rectangular after twisting. This is true for the shorter edge AB, and it is equally true for longer edge AC! Therefore along the corner edge AC the shear strain γ_{xy} is small or zero. Contrast this with the solid circular cylinder in (b). Here γ_{xy} at a point along AC equals the helix angle, α, because plane sections remain plane. Sketch (c) is a *reductio ad absurdum* which proves that certain sections have to warp.

10.4 THE ELEMENTS WANT TO WARP TOO

The formulation in FEMSKI uses the warping as the *only* nodal variable. We turn our back on St Venant's bubble, with few regrets. Apart from the loss of physical insight, what happens if some parts of the section are of different material? – or, as an extreme case, if the section contains holes? It can be done, but the solution is not obvious. The bubble has been the standard teaching analogy for three generations, we believe mainly because it permits dazzling displays of blackboard virtuosity. We do not wish to assassinate our more mathematical brethren; but the money spent today in actually *solving* such problems, in real engineering, far exceeds the money spent in teaching such 'advanced' material in the classroom.

In FEMSKI we take as a first approximation the case of a cylinder, Fig. 10.2(b). Whatever the shape of the section, we assume that plane sections remain plane, so that the shear strain is proportional to the distance from the 'centreline'. This leads to absurdities, as you have just seen. We then 'relax out' all such absurdities, by allowing each section to warp. We simply allow the sections to find the most comfortable positions – those that give the least strain energy. That is the whole story! An application of our general principle.†

10.5 SHEARING OF BEAMS

Why stop at torsion? Our beam, of uniform cross-section, may also be bent by an end-load. The bending stresses are easy, but the shearing due to the load is another matter. At no time have more than a very few people mastered the theory of the shearing of beams. Possibly you have learnt how to find the 'shear centre' for a beam made up of thin sheet metal. It is unlikely, however, that you understood what you were doing.

Under a shear load, the section warps, as it does in torsion. Using the same straightforward approach, a finite element package can discover the distribution of shear stresses over the cross-section, not only for a thin-walled beam, but for any section whatsoever. The shear centre follows. See Appendix III.

So does a remarkable and little-known phenomenon, the idea of 'principal shear areas' and of 'principal shear directions'. That is, regardless of the principal axes in bending, the additional shear deflection is in the direction of the applied load only if this lies along one of the 'principal shear directions'. The theory is similar, but, in general, these directions are not the same as the principal bending directions. All this follows, by 'calculating energies, drawing conclusions from them and trusting them'.

---

† Not quite the whole story in practice. Often where a beam is fixed, the warping is prevented. This may give very dangerous, very local stresses. The same sort of thing can happen in bending, section 10.5.

10.6 ALL FOR NOW

Our immediate task is complete. Yours had hardly begun. Remember this. Your practical experience already rivals that of many PhDs in the subject, which will be very useful in industry. But you must *not* become involved in finite element programming — not without a *lot* more study.

At the time of writing, this is an unconventional text. Much that you have learnt will puzzle the experts. Conversely, certain people will be scandalized at what you have *not* learnt. In your defence, we would argue that much of *that* is trivial or unnecessary. It is much more worthwhile to know about the hazards of finite elements, for example. Such topics have been neglected for too long. Our opinions.

Anyway, there is a bonus. You will now find that the conventional texts make new and *fascinating* reading. Best wishes.

FURTHER READING

Bathe, K.-J. and Wilson, E. L., *Numerical Methods in Finite Element Analysis*, Prentice-Hall (1976). A much-used book, for serious finite element programmers.

Hinton, E. and Owen, D. R. J., *Finite Element Programming*, Academic Press (1977). This teaches you to code the 8-node quadrilateral, with a simple front solution.

Irons, B. and Ahmad, S., *Techniques of Finite Elements,* Ellis Horwood (1980). The sequel to this book.

Zienkiewicz, O. C., *The Finite Element Method,* McGraw-Hill (third edition, 1977). The first book on finite elements.

Cat Leonora

Dog print Belle

Lucifer Rex

Bruce Irons   Nigel Shrive

Appendices

THE ELEMENT TYPES USED IN FEMSKI

Elements 1 to 6 can be triangles or quadrilaterals with or without midside nodes. Thus, the number of nodes, 3, 4, 6 or 8 specifies the shape and use (or not) of midside nodes.

(1) Membrane element with in-plane stresses, on an elastic foundation. Used for thermal conduction etc. but mainly included for instruction.
(2) Plane stress or plane strain element.
(3) Element for torsion of an axisymmetric body, e.g.: shafts of type 2. See Appendix II.
(4) Axisymmetric bodies with pressure, tension etc. but nothing to cause twisting: pressure vessels, discs etc.
(5) Bending of axisymmetric but non-uniform bodies.
(6) Twisting and shear deformation of shafts of uniform but irregular cross-section.
(7) Simple 2-node thin beam, in three dimensions.
(8) A hybrid 4-node quadrilateral: improved performance over type 2.
(9) A hybrid 8-node brick, also of high performance.
(10) Hybrid plate bending triangle and quadrilateral, 9 and 12 degrees of freedom.
(11) The Semiloof curved shell elements, a triangle with 24 degrees of freedom and a quadrilateral with 32 degrees of freedom.
(12) The special Semiloof beam element, designed to reinforce 11, 19 degrees of freedom.

APPENDIX II

INTRODUCTION to HYBRID ELEMENTS
THE MARKETPLACE

At the time of writing, commercial finite element packages compete mainly on the basis of versatility, ease of input, and availability of graphical output. One seldom hears anything about the element calculations themselves – apart from the recent campaign against the commercial plate bending elements that fail the patch test!

In the next few years, interest should return to questions of efficiency of alternative formulations. For example, when designers use a microcomputer, to model some detail, they will not want to use more than 6–10 elements. At this stage, users will be shamelessly manipulated, and callously brainwashed, and blinded with all kinds of pseudo-science, unless they have some understanding of 'hybrid elements', the favourite contenders. Yet it was very much as an afterthought that we added this appendix.

ENERGY

If you have read diligently through the book, and if you really believe in energy arguments, then you are ready to receive some physical insight into hybrid elements. This is probably enough: we can leave the technical details to the specialists. In a hybrid formulation,

(a) We postulate that within the element, the stresses are given in terms of M carefully chosen stress fields:

$\{\text{actual stress field}\} = A_1 \{\text{stress field 1}\}$

$$+ \ldots + A_M \{\text{stress field } M\} \quad (1)$$

The total stress field generates strain energy U over the whole element, which can be written

$$U = \tfrac{1}{2}[A_1 \ldots A_M][\mathbf{f}] \begin{bmatrix} A_1 \\ \vdots \\ A_M \end{bmatrix} \quad (2)$$

Because we like to regard the A_J as 'generalised forces', we call $[\mathbf{f}]$ a 'generalised flexibility matrix', by the same sort of argument that we used for stiffness matrices.

(b) The Stress fields 1 ... M are indeed exerting forces somewhere. These forces must interact with the outside world, via the nodal variables x_I, which connect for example to neighbouring elements. To link the x_I to the A, we require N ordinary shape functions, one for each nodal variable, as before.

What eventually emerges should be better than the shape function version itself.† For example, element 8 uses the same shape functions as element 2, but avoids the 'locking' trouble shown in Fig. 9.2, because we can omit any stress fields that would allow such undesirable responses. We must include the constant stress states however, and we must insist that the shape function version passes the patch test — that is all.

A simple example (Fig. A1) with only one degree of freedom, now introduces the subtle and clever technique, for replacing the 'bad' shape functions, which 'lock' etc., by 'good' shape functions, as implied by the chosen stress fields. In Fig. A1, X is a force, but A is not, in the usual sense. A multiplies into a 'unit' stress field, over a certain region of material. Let us put $A = 1$, so keeping the stresses constant by some magic, while we apply the virtual deflection $x = 1$.

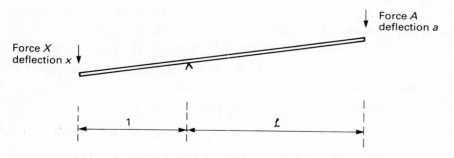

Fig. A1 – The leverage ratio is $\mathcal{L}:1$. It follows that force $X = \mathcal{L}A$ and that deflection $a = \mathcal{L}x$. If A and a are inaccessible, i.e. defined in some obscure manner, then virtual work can be used to find the effective $\mathcal{L}$, as in the derivation of hybrid elements.

† John Barlow has a beautiful way to introduce hybrids, at this stage. The total stress field (1) implies strains (h) say: the shape function version implies another, (s) say. With a fine mesh however, strains (s) ≈ strains (h): the differences should be slight, and the strain energy generated by {strains (h) − strains (s)} should be very small, even over the entire mesh. In fact, the hybrid formulation *minimises* this 'differential strain energy'. A hybrid explicitly tries to 'imitate' the shape functions, as closely as possible. That is, we choose the A_I in (1) by a 'least squares fit' – or rather, by the nearest structural equivalent.

The work done then equals the force **X**, and also equals the effective 'leverage ratio' $\mathcal{L}$, as required.

THE LEVERAGE MATRIX THEOREM

Virtual work, following Fig. A1, is also used in deriving the hybrid element, with many degrees of freedom. The generalised forces **A** perform the duties of the nodal forces **X**, through what we term the 'generalised leverage matrix' $[\mathcal{L}]$ ($M \times N$, for M stress fields and N nodal variables). The (I, J) term of $[\mathcal{L}]$ is the force along x_I due to $A_J = 1$. We calculate $\mathcal{L}(I, J)$ from the virtual work done on the Jth stress field, by the strains that would be induced by the Ith shape function ($x_I = 1$) at each integration point. Thus the virtual work done by all the x_I on the total stress field (1), artificially held constant as we move the nodes, may be written

$$W = \mathbf{x}^T \mathcal{L} \mathbf{A}$$

$$= \mathbf{x}^T \mathbf{X} \tag{3}$$

where each force **X** also remains constant while we impose the deflections **x**. Alternatively,

$$W = \mathbf{x}^T \mathcal{L} \mathbf{A}$$

$$= \mathbf{a}^T \mathbf{A} \tag{4}$$

where $\mathbf{a} = \mathbf{f}\mathbf{A}$ comprises the 'generalised displacements' corresponding to **A**.

Equation (3) suggests that

$$\mathbf{X} = \mathcal{L}\mathbf{A}, \tag{5}$$

and equation (4) suggests that

$$\mathbf{a}^T = \mathbf{x}^T \mathcal{L}, \quad \text{giving} \quad \mathbf{a} = \mathcal{L}^T \mathbf{x} \dots \tag{6}$$

In fact, (5) follows directly from virtual work: $\mathcal{L}(I, J)$ is the force along x_I due to the unit stress field J. But (6) is more challenging. If $[\mathcal{L}]$ was scalar, a simple leverage ratio, the transition from (5) to (6) would be trivial, as in Fig. A1. But $[\mathcal{L}]$ is a rectangular matrix, and (6) is decidedly nontrivial. It is called a 'theorem', with several different names. We do not prove it. Your finite element salesman is unlikely to want to argue about it.*

Accepting theorem (6), the remaining steps are painless:

$$U = \tfrac{1}{2}\,\mathbf{A}^T \mathbf{f} \mathbf{A}, \text{ from (1)}$$

$$= \tfrac{1}{2}\,\mathbf{a}^T \mathbf{f}^{-1} \mathbf{a}, \text{ because } \mathbf{a} = \mathbf{f}\mathbf{A}$$

$$= \tfrac{1}{2}\,\mathbf{x}^T \mathcal{L} \mathbf{f}^{-1} \mathcal{L}^T \mathbf{x}, \text{ substituting } \mathbf{a} \text{ from (6)}$$

which gives

$$\mathbf{K} = \mathcal{L}\mathbf{f}^{-1}\mathcal{L}^T \tag{7}$$

COMPLICATIONS

The only difficulty in (7) is in understanding, and in some cases actually calculating, the leverage matrix $[\mathcal{L}]$. For example, in plate bending there is a less happy theorem: if you have the deflection and two slopes as the variables at each corner, then the only slope-conforming and patch-passing shape functions possible must have weird properties, and you cannot use simple polynomials†. A pyramid with quadrilateral base is equally unpleasant.

In such cases, we can forego the luxury of shape functions, and make do with functions that define the responses only over the element boundaries, including the slopes in plate bending. Thus the virtual work contributions to $[\mathcal{L}]$ must be computed, term by term, edge by edge or face by face, which is much more tedious. Also, to ensure that the answers are exactly the same as those that would result from overall shape functions, we must now insist that the stress fields are equilibrating everywhere, so that all the virtual work is done at the boundaries – not that many people would be tempted to risk the alternative, in any case!

NUMERICAL EXAMPLE

The thin beam shown in Fig. A2 merely illustrates the technique. To find the generalised flexibility matrix, consider the strain energy:

---

*In the literature, you will probably find H for F, and T for $\mathcal{L}$. You can read more about our approach in 'Techniques of Finite Elements'. Theorem (6) emerges directly from the Lagrange multiplier equations to minimize U subject to constraints (5): the Lagrange multipliers are recognised as the nodal deflections x_I by Section 27.23. Chapter 15 is a mess, and contains far too much irrelevant detail.

†If you are interested, see 'Techniques of Finite Elements', Sections 16.3.

$$U = \frac{1}{2\,EI} \int_{-b}^{b} [A_1 A_2] \begin{bmatrix} 1 \\ x \end{bmatrix} [1, x] \begin{bmatrix} A_1 \\ A_2 \end{bmatrix} \mathrm{d}x$$

$$= \frac{1}{2\,EI} [A_1 A_2] \begin{bmatrix} 2b & 0 \\ 0 & \frac{2b^3}{3} \end{bmatrix} \begin{bmatrix} A_1 \\ A_2 \end{bmatrix}$$

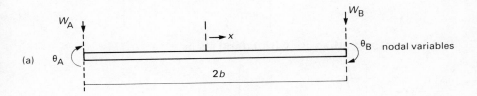

(a)

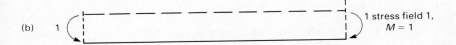

(b)

(c)

Fig. A2 – The nodal variables, and the constant and linear bending moment fields, for a uniform thin beam.

Thus $\quad \mathbf{f} = \frac{1}{EI} \begin{bmatrix} 2b & 0 \\ 0 & \frac{2b^3}{3} \end{bmatrix}$

$$\mathbf{f}^{-1} = \frac{EI}{2b^3} \begin{bmatrix} b^2 & 0 \\ 0 & 3 \end{bmatrix}$$

$$\mathcal{L}^{\mathrm{T}} = \begin{bmatrix} 0 & -0 & 0 & 1 \\ 1 & b & -1 & b \end{bmatrix}$$

For example, the second row of $[\mathcal{L}^{\mathrm{T}}]$ gives the work done directly by the end moments and shear force of the second stress field, in

Fig. A2(c), on the arrows representing the variables, taken in order $[w_A \ \theta_A \ w_B \ \theta_B]$ in (a). Thus

$$\mathcal{L}f^{-1}\mathcal{L}^T = \mathbf{K} = \frac{EI}{2b^3}\begin{bmatrix} 3 & 3b & -3 & 3b \\ & 4b^2 & -3b & 2b^2 \\ & & 3 & -3b \\ \text{symm.} & & & 4b^2 \end{bmatrix}$$

More interesting is the case with lateral shear:

$$[\mathbf{f}] = \begin{bmatrix} \dfrac{2b}{EI} & 0 \\ 0 & \dfrac{2b^3}{3EI} + \dfrac{2b}{GA} \end{bmatrix}$$

etc., where A is the effective shear area — see 'Structural Analysis', Ghali and Neville, p. 401, Intext (1972).

APPENDIX III

SHEARING OF BEAMS – ENERGY APPROACH
– a brief introduction to principal shear directions

Imagine a uniform beam extending from $z = 0$ (fixed) to $z = \ell$, with a torsional moment M_z, and a load V_x, V_y acting at $z = \ell$, at the point in the cross-section that we choose as origin x and y. We can calculate M_x and M_y at any other section z: V_x, V_y and M_z do not change with z. The bending and the torsion are easy anyway, but we could treat them by energy, if we so wished. That is, we could use thin beam theory, assuming that $u = u_o z^2(\frac{1}{2}\ell - \frac{1}{6}z)/\ell^3$, $v = v_o z^2(\frac{1}{2}\ell - \frac{1}{6}z)/\ell^3$, $\theta_z = \theta_o(\ell - z)/z$, with u_o, v_o and θ_o as the values at $z = \ell$. (We know the form of the answers, regardless of the section properties.) Writing the energy of the beam in the form

$$\tfrac{1}{2}[u_o \ v_o \ \theta_o] \ [\mathbf{K}] \begin{bmatrix} u_o \\ v_o \\ \theta_o \end{bmatrix}$$

we recognise $[\mathbf{K}]$ as a stiffness matrix. Inverting it, we have $[\mathbf{K}]^{-1} = [\mathbf{f}_1]$, the flexibility matrix without shearing. The eigenvectors of $[\mathbf{f}_1]$ – the u and v part proportional to ℓ^3 – give the principal directions in bending, i.e. the load directions which would give displacements in the same direction as the load.

With a thick beam there is warping due to the shear force, which gives a small extra flexibility $[\mathbf{f}_2]$, proportional to ℓ:

$$[\mathbf{f}] = [\mathbf{f}_1] + [\mathbf{f}_2]$$

In this case the warping is not just a matter of relaxing. The warping is the same in every section, so that because σ_z due to bending varies with z, the warping of any fibre does work on the bending stress. On this basis, we can find the $[\mathbf{K}_2]$ in u_o and v_o, and hence find $[\mathbf{f}_2]$ which has eigenvectors giving the directions in which the additional shear deflection will be in the same direction as the force V. These directions are the principal shear directions, and they are not necessarily the same as the principal bending directions. This is probably of no practical importance, because normally the shear deflections are small, but you never know for sure. ...

Index

In principle, a good index should enable you, not only to dip casually into a book that you never intend to read seriously, but also to jump around in a book that you intend to master, so as to learn as quickly as you can. In other words, two people with different backgrounds could read the same book in quite a different order, as with the 'programmed texts' and teaching machines that were in vogue around 1970. It is a free world – you should be able to – and this is an implied criticism of certain undergraduate texts. Whatever you feel about the future of education, please put this index to the test sometimes, and let us know about your experiences. Many people think that the index is the most important part of a textbook. We do.

DATE DUE

DATE DUE			
AUG 1 1989			
JUL 1990			

DEMCO 38-297